Stress at Work

Management and Prevention

Stress at Work

Management and Prevention

Jeremy Stranks

ELSEVIER
BUTTERWORTH
HEINEMANN

AMSTERDAM BOSTON HEIDELBERG LONDON NEW YORK OXFORD
PARIS SAN DIEGO SAN FRANCISCO SINGAPORE SYDNEY TOKYO

Elsevier Butterworth-Heinemann
Linacre House, Jordan Hill, Oxford OX2 8DP
30 Corporate Drive, Burlington, MA 01803

First published 2005

British Library Cataloguing in Publication Data
A catalogue record for this book is available from the British Library

Library of Congress Cataloguing in Publication Data
A catalogue record for this book is available from the Library of Congress

ISBN 0 7506 6542 4

For information on all Elsevier Butterworth-Heinemann publications visit our website at http://books.elsevier.com

Typeset by Charon Tec Pvt. Ltd, Chennai, India
www.charontec.com
Printed and bound in Great Britain by Biddles Ltd, King's Lynn, Norfolk

Contents

Preface

People at work worry about all sorts of things – increasing competition for jobs, globalization, terrorism, 'rationalization' of the organization's operations, looking after ageing parents and relatives, the threat of redundancy, annual appraisals, new technology, outsourcing of jobs to India and other Third World countries together with increased demands by employers for higher productivity. Moreover, they may be put under excessive pressure at certain times, for example, to meet sales targets, attend meetings on time, learn and follow new procedures and fit in with changes in the organization's culture. This can result in varying levels of stress. According to the Health and Safety Executive, workplace stress is now the fastest growing cause of absence from work.

What sort of employer are you? When your employees complain of stressful conditions at work, do you reply with the old maxim 'If you can't stand the heat, get out of the kitchen!'? The days when such a response from employers was common are over. Employers now need to get to grips with a range of policies and procedures to deal with stress at work.

What is important is that the poor standards of performance by many employees due to the effects of stress at work represent a substantial financial loss to their organizations and the British economy. Moreover, recent cases in the civil courts, and the greater attention now being paid to the subject of stress at work by the enforcement agencies, means that employers need to consider stress in the workplace and the measures they must take to prevent employees suffering stress arising from their work. It is not uncommon for six figure sums to be awarded as damages in civil claims for stress-induced injury.

This book has been written as a guide for managers. It should enable them to understand the meaning of stress, the causes of stress, human responses to stress and aspects of behaviour which are significant in this area. In particular, employers need to manage stress by incorporating stress protection for their employees into their management systems.

The book incorporates a number of important features, including a stress audit, the recent Court of Appeal general guidelines with respect to civil claims for stress-related ill health, measures necessary with respect to bullying and harassment and procedures for bringing stress management into operating procedures.

I would like to thank Dr Jacques Tamin of Interact Health Management for contributing the work-related stress (WRS) risk assessments.

Jeremy Stranks
August 2004

1

Introduction to stress

The recent civil court decision in which a senior social worker was awarded £175 000 compensation against his local authority employer for allowing him to work to the point of breakdown raises the question as to whether claims for stress at work will be the significant legal issue of the next decade.

Other claims have followed, including the claim involving a ticket collector who received £375 000 in damages for post-traumatic stress disorder as a result of the King's Cross station fire.

Stress at work, and the potential for stress-induced ill health, has become a topical subject with many people. Furthermore, most people can describe stressful events and circumstances at work. Inefficient management, lack of decision-making by management, excessive working hours, uncertainty as to future employment prospects and the pressure of the job are some of the causes of stress described by employees.

1.1 What is stress?

'Stress' is a word which is rarely clearly understood and there is no single definition of the term. It means different things to different people. Indeed, almost anything anyone can think of, pleasant or unpleasant, has been described as a source of stress, such as getting married, being made redundant, getting older, getting a job, too much or too little work, solitary confinement or exposure to excessive noise.

Stress can be defined in many ways, thus:

- The common response to attack (Selye, 1936);
- Any influence that disturbs the natural equilibrium of the living body;
- Some taxation of the body's resources in order to respond to some environmental circumstance;
- The common response to environmental change;

- A psychological response which follows failure to cope with problems;
- A feeling of sustained anxiety which, over a period of time, leads to disease;
- The non-specific response of the body to any demands made upon it.

The CBI defines stress as that which arises when the pressures placed upon an individual exceed the perceived capacity of that individual to cope.

According to the TUC, stress occurs where demands made on individuals do not match the resources available or meet the individual's needs and motivation. Stress will arise if the workload is too large for the number of workers and time available. Equally, a boring or repetitive task which does not use the potential skills and experience of some individuals will cause them stress.

The Health and Safety Executive (HSE) (1995) defined work stress as 'pressure and extreme demands placed on a person beyond his ability to cope'. In 1999, the Health and Safety Commission (HSC) stated that 'stress is the reaction that people have to excessive pressures or other types of demand placed upon them'.

According to Cox (1993), 'stress is now understood as a psychological state that results from people's perceptions of an imbalance between job demands and their abilities to cope with those demands'.

A further definition is 'work stress is a psychological state which can cause an individual to behave dysfunctionally at work and results from people's response to an imbalance between job demands and their abilities to cope'.

Fundamentally, workplace stress arises when people try to cope with tasks, responsibilities or other forms of pressure connected with their jobs, but encounter difficulty, strain, anxiety and worry in endeavouring to cope.

1.2 Defining stress

A consideration of the above definitions of 'stress' produces a number of features of stress and the stress response, for example, disturbance of the natural equilibrium, taxation of the body's resources, failure to cope, sustained anxiety, a non-specific response, pressure and extreme demands and imbalance between job demands and coping ability.

Fundamentally, a stressor (or source of stress) produces stress which, in turn, produces a stress response on the part of the individual. No two people respond to the same stressor in the same way or to the same extent. What is important is that, if people are going to cope satisfactorily with the stress in their lives, they must recognize:

- The existence of stress;
- Their personal stress response, such as insomnia or digestive disorder;
- Those events or circumstances which produce that stress response, such as dealing with aggressive clients, preparing to go on holiday or disciplining employees;
- Their own personal coping strategy, such as relaxation therapy.

1.3 Degradation of human performance

Human performance is directly affected by the environment in which people work and sound levels of working environment promote optimum levels of performance. Many factors influence the human system and performance can degrade as a result of a wide range of stressors, and in some cases the system breaks down.

Degradation of performance is particularly associated with the following stressors.

1.3.1 Diurnal (circadian) rhythm

Body rhythms tend to follow a cyclical pattern linked to the 24-h light–dark cycle and sleeping–waking cycle, that is diurnal rhythm. Interruptions in this rhythm, as experienced by, for example, casual workers, shift workers and night workers, can cause stress on operators resulting in reduced operational performance as much as 10 per cent below average performance.

In the case of night workers, adjustment may take place after 2–3 days and goes on increasing up to a period of approximately 14 days provided that the individual continues both to live and work on a night-time schedule, and does not return to normal daytime living at weekends.

Rotating shift patterns, for example, a week on night work followed by a week on day work, or the operation of 12-h shifts rotating from, for instance 6 a.m. to 6 p.m., noon to midnight and 6 p.m. to 6 a.m. on different weeks, can result in high levels of stress on operators and their families.

1.3.2 Fatigue

Fatigue commonly results from working excessive hours without rest breaks and adequate periods of sleep.

1.3.3 Lack of motivation

Where there is no stimulation from management in terms of performance targets and the rewarding of employees for achieving these targets, employees rapidly become demotivated and their performance deteriorates.

1.3.4 Lack of stimulation

Many jobs are boring, repetitive and demotivating resulting in a lowered level of arousal on the part of operators. Stimulation of performance can be achieved by job rotation, productivity bonus schemes (provided the rewards are seen to be fair to all concerned), working in small teams and, in certain cases, counselling of employees in an endeavour to reduce stress.

1.3.5 Stress

As stated above, a stressor causes stress. Stress is commonly associated with how well or badly people cope with changes in their lives – at home, within the family, at work or in social situations. As will be seen in Chapter 2, the causes are diverse, but include:

- **Environmental stressors**, such as those arising from extremes of temperature and humidity, inadequate lighting and ventilation, noise and vibration and the presence of airborne contaminants, such as dusts, fumes and gases;
- **Occupational stressors**, associated with too much or too little work, over-promotion or under-promotion, conflicting job demands, incompetent superiors, working excessive hours and interactions between work and family commitments; and
- **Social stressors**, namely those stressors associated with family life, marital relationships, bereavement, that is, the everyday problems of coping with life.

1.4 The evidence of stress

Research in the 1990s by Professor Cox of Nottingham University led to much of the HSE's current guidance on the subject. Following an independent review of the literature, Professor Cox indicated that there was a reasonable consensus from the literature on psychosocial hazards (or stressors) arising from work which may be experienced as stressful or otherwise, and that these stressors may carry the potential for harm. According to the research there are nine characteristics of jobs, work environments and organizations which were identified as being associated with the feeling of stress and which could damage or impair health.

These characteristics are of two types, context or setting and nature:

1. The context or setting in which the work takes place, i.e.:
   - organizational function and culture
   - career development
   - decision latitude/control
   - role in organization
   - interpersonal relationships
   - the work/home interface.
2. The content or 'nature' of the job itself, in particular:
   - task design
   - workload or work pace
   - work schedule.

Further research released by the HSE gives an indication of the scale of the problem of injuries which are stress-related. In the report *The Scale of Occupational*

Stress: The Bristol Stress and Health at Work Study CRR 265/2000 (Smith et al., 2000), it was estimated that there are 5 million workers suffering from high levels of stress at work. Important outcomes of this study were:

- Approximately one in five workers reported high levels of stress arising from work.
- There was an association between high levels of reported stress and specific job factors such as excessive workloads or lack of managerial support.
- There was an association between high levels of reported stress and certain aspects of ill health, such as poor mental health and back pain, together with certain health-related activities such as smoking and excessive alcohol intake.

What came out of this study is that stress is now a foreseeable cause of ill health and that employers need to take this factor into account when considering the means for reducing the running costs of the undertaking.

1.5 Stress as opposed to pressure

Not all stress, however, is bad for people. Most people need a certain level of positive stress or pressure in order to perform well the tasks allotted to them. Some people are capable of dealing with very high levels of positive pressure. This is the classic fight response or 'butterfly feeling' that people encounter before sitting an examination, running a race or attending a job interview.

Positive stress is one of the outcomes of competent management and mature leadership where everyone works together and their efforts are valued and supported. It enhances well-being and can be harnessed to improve overall performance and fuel achievement.

It is the negative stress, or distress, such as that arising from having to meet set deadlines or delegate responsibility, commonly leading to ill health, that needs to be considered by employers as part of a stress management strategy. It may be the result of a bullying culture within the organization where threat, coercion and fear substitute for non-existent management skills. With this sort of culture, employees have to work twice as hard to achieve half as much to compensate for the dysfunctional and inefficient management. Negative stress diminishes quality of life and causes injury to health resulting in a range of stress-related symptoms.

1.6 The cost of stress

In recent years organizations such as the CBI, TUC, Department of Health and the HSE, together with an increasing number of both large and small employers, have expressed concern about the increasing costs of stress at work, not only in human and financial terms, but to the national economy generally.

Early studies into the cost of stress at work identified a number of important points with respect to the cost of stress at work.

- Stress is said to cost British industry approximately 3 per cent of the gross national product.
- Stress-related costs amount to more than 10 times the cost of all industrial disputes.
- Stress-related illness directly causes the loss of 40 million working days each year.

The cost of replacing an employee who is underperforming owing to stress is between 50 and 90 per cent of his annual salary (Personnel Management, Factsheet 7, July 1988). More recently, in the HSE report *Mental Health and Stress in the Workplace: A Guide for Employers* (1996), it was estimated that 360 million working days were lost annually in the UK at a cost of £8 billion, and that half of these absences were stress-related. Moreover, the pilot results of a national survey into stress at work, originally launched in 1997 by the University of Bristol on behalf of the HSE, revealed that every day 270 000 people are absent from work with a stress-related illness.

The CBI estimates that stress and stress-related illness cost UK industry and taxpayers £12 billion each year. The UK Department of Health state that 3.6 per cent of national average salary budget is paid to employees off sick with stress. In fact, stress is now officially the prime cause of sickness absence, although 20 per cent of employers still do not regard stress as a health issue.

1.7 The response of the courts to stress

Employers should not only be concerned about the problems of reduced productivity and absenteeism associated with stress, however. What is of particular concern is the dramatically increased attention to stress being given by the courts and enforcement agencies. In fact, stress-related injury has been described as 'the civil claim of the millennium'. *Walker v Northumberland County Council* (1995) was the landmark case in the civil courts focusing attention on the subject of work-related stress for the first time. In this case, Walker, who was a social worker, suffered two nervous breakdowns due to stress and overwork. He subsequently sued his employers, Northumberland County Council, and was awarded £175 000 in damages. The total costs of this case, however, were nearly £500 000 when legal costs, sick pay and pension were taken into account.

Since this case, there has been an approximately 90 per cent increase in civil claims for mental and physiological damage. Further stress-related claims are dealt with in Chapter 8.

It is significant that the HSE has taken the issue of stress at work on board in recent years resulting in a range of publications on the subject directed at employers with a view to reducing stress at work. The criminal implications of stress at work could be expensive for employers in future years in terms of fines in the criminal courts. A number of questions need to be asked in this case with respect to the criminal liability of employers.

- What will be the response of the enforcement agencies to complaints from employees of stress at work?
- Is it likely that an employer could be served with an improvement notice under the Health and Safety at Work Act to, for example, install a stress management programme in his organization?

- Are employers likely to be prosecuted where there is evidence of stress amongst employees?
- Who will be the experts in determining, firstly, whether an employee is suffering stress-related ill health and, secondly, predicting the short-, medium- and long-term effects of that stress?
- Is it conceivable that, in years to come, stress at work regulations will be brought out laying down requirements and procedures for employers on this matter?

1.8 The physiology of stress

Stress could be defined simply as the rate of wear and tear on the body systems caused by life. The acknowledged father of stress research, Dr Hans Selye, a Vienna-born endocrinologist of the University of Montreal, in his book *The Stress of Life* corrected several notions relating to stress, in particular:

- Stress is not nervous tension.
- Stress is not the discharge of hormones from the adrenal glands. The common association of adrenalin with stress is not totally false, but the two are only indirectly associated.
- Stress is not simply the influence of some negative occurrence. Stress can be caused by quite ordinary and even positive events, such as a passionate kiss.
- Stress is not an entirely bad event. We all need a certain amount of stimulation in life and most people can thrive on some forms of stress.
- Stress does not cause the body's alarm reaction, which is the most common misuse of the expression. What causes the stress reaction or response is a stressor.

A number of common factors emerge from the definitions of stress outlined earlier and the above comments. Fundamentally, stress is a state manifested by a specific syndrome of biological events. Specific changes occur in the biological system, but they are caused by such a variety of agents that stress is, of necessity, non-specifically induced.

Some stress response, however, will result from any stimulus. Quite simply, a stressor produces stress. Stressors may be of an environmental nature such as extremes of temperature and lighting, noise and vibration (environmental stressors). Stress may be induced by isolation, rejection, change within the organization or the feeling that one has been badly treated (social stressors). Thirdly, stress can be viewed as a general overloading of the body systems (distress).

1.8.1 The autonomic system

Stress has a direct association with the autonomic system which controls an individual's physiological and psychological responses. This is the flight or fight syndrome, characterized by two sets of nerves, the sympathetic and parasympathetic, which are responsible for the automatic and unconscious regulation of body function.

Table 1.1 Sympathetic and parasympathetic balance

Parasympathetic state	Sympathetic state
Eyes closed	Eyes open
Pupils small	Pupils enlarged
Nasal mucus increased	Nasal mucus decreased
Saliva produced	Dry mouth
Breathing slow	Breathing rapid
Heart rate slow	Heart rate rapid
Heart output decreased	Heart output increased
Surface blood vessels dilated	Surface blood vessels constricted
Skin hairs normal	Skin hairs erect (goose flesh)
Dry skin	Sweating
Digestion increased	Digestion slowed
Muscles relaxed	Muscles tense
Slow metabolism	Increased metabolism

The sympathetic system is concerned with answering the body's call to fight, i.e. increased heart rate, more blood to the organs, stimulation of sweat glands and the tiny muscles at the roots of the hairs, dilation of the pupils, suppression of the digestive organs, accompanied by the release of adrenalin and noradrenalin.

The parasympathetic system, on the other hand, is responsible for emotions and protection of the body, which have their physical expression in reflexes, such as widening of the pupils, sweating, quickened pulse, blushing, blanching, digestive disturbance, etc.

The balance between the sympathetic and parasympathetic systems is shown in Table 1.1.

1.8.2 The general adaptation syndrome

Stress is a mobilization of the body's defences, an ancient biochemical survival mechanism perfected during the evolutionary process, allowing human beings to adapt to threatening circumstances. In 1936, Selye defined this 'general adaptation syndrome' which comprises three stages.

1. **The alarm reaction stage**. This is typified by receiving a shock, at the time when the body's defences are down followed by a counter-shock, when the defences are raised. In physiological terms, once a stressor is recognized, the brain sends out a biochemical messenger to the pituitary gland which secretes adrenocorticotrophic hormone (ACTH). ACTH causes the adrenal glands to secrete corticoids, such as adrenalin. The result is a general call to arms of the body's systems.
2. **The resistance stage**. This stage is concerned with two responses. The body will either resist the stressor or adapt to the effects of the stressor. It is the opposite of the alarm reaction stage, whose characteristic physiology fades and disperses as the organism adapts to the derangement caused by the stressor.
3. **The exhaustion stage**. If the stressor continues to act on the body, however, this acquired adaptation is eventually lost and a state of overloading is reached.

Table 1.2 The stress response

The response	What happens	The effect
Flight or fight	Red alert, body and brain prepare for action; extra energy released	Response to danger, meet it and return equilibrium
Secondary	Fats, sugars and corticosteroids released for more energy	Unless extra fats etc. used up, then third stage moved into
Exhaustion	Energy stores used up	Serious illness leading to death

> The symptoms of the initial alarm reaction stage return and, if the stress is unduly prolonged, the wear and tear will result in damage to a local area or the death of the organism as a whole.

The three stages of the stress response can be summarized as shown in Table 1.2.

1.8.3 Selye's model

Selye's model illustrating the general adaptation syndrome is shown in Figure 1.1. This model shows the individual surrounded by a variety of stressors. His response to these stressors is affected by factors such as his strength of constitution, psychological strength, degree of control over the situation and how he actually perceives the potentially stressful event.

The effect of these stressors is to require some form of general adaptation by the individual. Here the situation can go one of two ways. If the individual adapts unsuccessfully, this leads to further wear and tear on the mind and body, general weakness and stress-related illness. This, in turn leads to increased vulnerability to further stressors in his life. Successful adaptation, on the other hand, leads to growth, happiness, security and strength, with greater resistance to further stressors.

No two people respond to the same stressor in the same way. However, in the majority of cases, exposure to a stressor will produce some form of personal stress response. This stress response could be digestive disorder, irritability or raised heart rate. Insomnia is a classic manifestation of stress.

Similarly, different people are affected by different stressors, such as boredom at work, the introduction of new technology or their lack of career development. For some people, stress may be created by trying to satisfy the demands of work and, at the same time, the demands of a young family, the classic 'home–work interface'.

What comes out of Selye's model is the fact that people need, firstly, to recognize those situations, circumstances and events that create a specific stress response in themselves, such as digestive disorders and increased respiration rate and, secondly, to develop their own personal strategies for coping with the particular stressors.

Fundamentally, this 'flight or fight' mechanism or stress response is designed for responding to physical danger, such as being chased by a lion. However, in the work

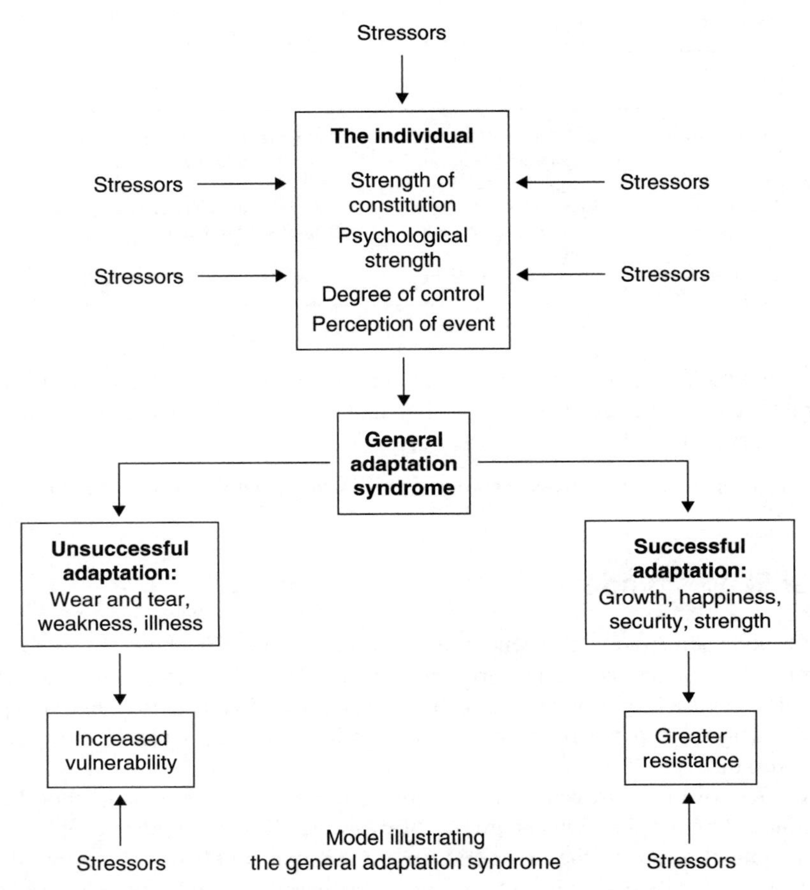

Figure 1.1 General adaptation syndrome.

situation, it is more likely to be set in motion by psychological danger, such as bullying or harassment by work colleagues or verbal and physical abuse. The stress response can also be activated in anticipation of adverse situations, such as loss of job, inability to pay a debt, being stopped by the police for exceeding a speed limit or being caught up in a road rage incident.

As stated previously no two people respond to the same stressor in the same way, and respond with differing degrees of stress. However, there are a number of factors which determine the level to which an individual will feel stressed:

- **Control**: A person will demonstrate stress to the extent to which they perceive they are not in control of a stressor. Generally, employees have no control over their employers.
- **Predictability**: A person will feel stressed due to the extent of his inability to predict the behaviour or occurrence of a stressor. Bullies, for example, are notoriously unpredictable in terms of what they are going to do next.

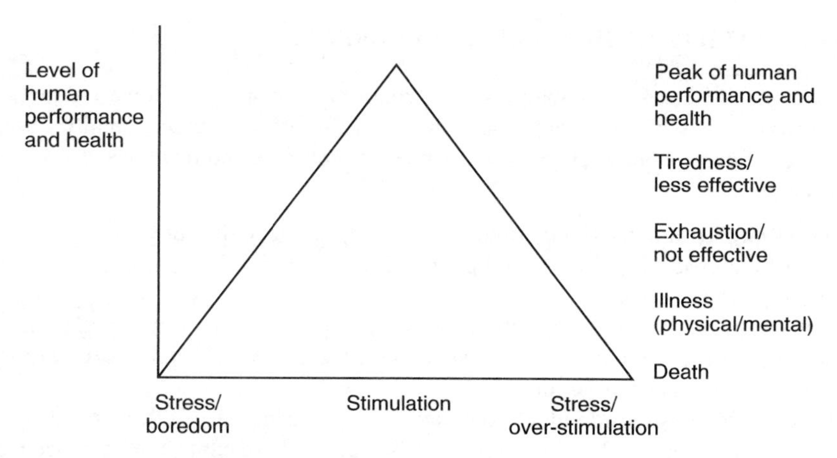

Figure 1.2 An adaptation of the human performance curve.

- **Expectation**: An individual will suffer stress to the extent to which he perceives his circumstances are still not getting better and will not get better. In this case, a bullying situation will inevitably deteriorate further.
- **Support**: A person will feel stressed to the extent to which he lacks support, including colleagues at work, managers, trade union representatives, family, friends, people in authority, his doctor and official organizations.

1.9 A model of human performance and stress

The model in Figure 1.2 demonstrates the significance of the 'human performance curve'. At the peak of the curve, the individual has reached his peak of performance and health. However, as a result of stress-related boredom or overstimulation, the level of human performance can deteriorate through a number of stages whereby the individual experiences tiredness and becomes less effective. A further stage is reached indicating exhaustion and a loss of effectiveness. This can eventually result in both physical and mental illness leading, in some cases, to death.

This model highlights lack of challenge as having similar effects to overstimulation, and that these effects can be progressive. It can also be used to raise the point that early recognition of the effects of stress can enable the individual to prevent their progression into ineffectiveness and ill health.

1.10 The effects of stress

Stress can have a significant effect, both on the individual and the organization.

1.10.1 Effects of stress on the individual

No two people necessarily manifest the same stress response. However, many of the outward signs of stress are readily recognizable. Stress fundamentally initiates a number of changes in body processes which are complex and involve several levels, such as:

1. **Emotional**: characterized by tiredness, anxiety and lack of motivation;
2. **Cognitive**: resulting in increased potential for error and, in some cases, accidents arising through error;
3. **Behavioural**: changes in behaviour resulting in poor or deteriorating relationships with colleagues, irritability, indecisiveness, absenteeism, smoking, excessive eating and alcohol consumption;
4. **Psychological**: the individual complains of increasing ill health associated with headaches, general aches and pains, and dizziness. These contribute to raised blood pressure, heart disease, a reduced resistance to infection, skin conditions and digestive disorders.

The HSE (2002) contract research report *Work Environment, Alcohol Consumption and Ill Health, The Whitehall II Study CRR 422/2002* confirmed that a stressful working environment can lead to coronary heart disease, in most cases associated with ever-increasing job demands, poor levels of actual control over the job and an imbalance between the efforts made and the reward received. However, these effects were not necessarily related to conventional risk factors, such as smoking, high blood pressure and being overweight. Broadly, when people are subjected to changing workloads, resulting in higher demands being imposed, less direct control over the job and reduced support from management, their mental health deteriorated.

The report from the *Whitehall II Study* on the health of over 10 000 civil servants in the UK examined the influence of:

1. Job demands
2. The amount of influence employees have over how they do their work (job control)
3. The level of support from managers and colleagues
4. The effect on physical health of an imbalance in the effort people put into work and the rewards arising from same.

These factors are directly related to how stressful people find their work and could be applied to many other groups of workers. Previous reports have linked working conditions with heart disease, but this report gives a more clear and accurate assessment as, in this case, the participants' reports of heart disease were verified against their medical records.

1.10.2 Effects of stress on job performance

For people to perform well, they need interesting work, good working conditions, the chance to partake in the social surroundings of work and to feel valued.

Stressful work situations arising from, for example, the need for boring or repetitive work patterns, such as assembly work, poor physical working environments, isolated working situations, inadequate opportunities for communication between colleagues and continuous harassment from managers to meet deadlines can have direct effects on job performance. In particular, where people feel their contribution to the organization's success is undervalued, this can result in missed deadlines, poor productivity, ineffective decision-making by line managers and, in many cases, poor time keeping and absenteeism.

1.10.3 Effects of stress on the organization

Attitudes to stress amongst managers at all levels vary considerably. In some organizations, the culture can only be defined as 'aggressive'. Employees who complain about stress caused by excessive workloads may be greeted with the classic 'If you can't stand the heat, get out of the kitchen!' response from their immediate manager.

In many organizations, junior managers are expected to work long hours and undertake a range of projects and assignments in order to prove their worth to the organization. Furthermore, job and career reviews (appraisals) are intended to provide guidance to junior managers from senior management, to review current progress and to agree objectives, which are measurable and achievable, for future performance. Many job and career reviews, however, are badly conducted and can be stressful for employees, frequently resulting in stress arising from a feeling of unfairness, lack of understanding by their immediate manager and resentment.

Examples of how stress can affect an organization include:

- Increased complaints from clients;
- Employees losing commitment to the success of the organization;
- Increased accidents;
- Increased staff turnover;
- Increased levels of absenteeism;
- Reduced performance by the workforce; and
- A substantial increase in civil claims for stress-induced injury resulting in increased employers' liability insurance premiums.

Well-informed managers should recognize the signs of stress amongst employees at all levels. Failure to do this can have lasting adverse effects on the business, including low motivation, increased absenteeism, reduced productivity, faulty decision-making, poor industrial relations and reduced efficiency.

1.11 Occupational groups

Certain occupations have been shown to be more stressful than others. The HSE report (2002) *The Scale of Occupational Stress: A Further Analysis of the Impact of Demographic Factors and the Type of Job CRR 311*, indicates that the occupational groups reporting high levels of occupational stress are, in order, teachers, nurses, managers, professional persons, other people involved in education and welfare (which includes social workers), road transport personnel and those involved in security operations, which includes police and prison officers.

Out of the above groups at least one in five reported high levels of stress. In the case of teachers, the figure was two in five. This research also showed that:

1. Full-time workers reported being more stressed than those in part-time work.
2. Those in managerial and technical posts reported being highly stressed, along with those educated to degree level and those earning in excess of £20 000 a year.
3. There was evidence of a racial element, although the numbers involved were small. Here non-white workers reported higher levels of stress than white workers.
4. There was little evidence of any significance of gender as there was little difference in reported levels between men and women.

1.12 Conclusion

Most people need a certain level of stress in order to perform well (positive stress). However, the benefits can rapidly turn to negative stress as a result of work overload situations, conflict situations in the workplace or the feeling of insecurity as a result of organizational changes.

How well or how badly people adapt to changes in their lives is a significant factor in the consideration of stress. As Selye demonstrated, for some people unsuccessful adaptation to change can have serious health effects. For others, successful adaptation brings growth, greater happiness and increased resistance to stress. In order to survive stressful events in their lives, people need to be more aware of stress, their personal stress responses and of strategies for coping with stress.

What is important is that organizations can no longer ignore evidence of stress amongst employees at all levels. Systems for managing stress and the strategies necessary for reducing same are covered in Chapters 6 and 7.

Questions to ask yourself after reading this chapter

- What is meant by 'stress'?
- What is the flight or fight response?
- How does the body's autonomic system operate?
- What are the various stages of the general adaptation syndrome?
- What are the effects of stress on the individual?
- Which occupational groups are most commonly exposed to stress?
- What is meant by positive stress?
- What are the effects of stress on the behavioural processes of people?
- What is the human performance curve?
- What are the functions of the body's sympathetic and parasympathetic systems?

Key points – implications for employers

- 'Stress' is defined in many ways. The HSE define stress as 'pressure and extreme demands placed on a person beyond his ability to cope'.
- Not all stress is bad for people. Most people need a certain amount of positive stress or 'pressure' in order to perform well the tasks allotted to them.
- It is estimated that stress costs British industry approximately 3 per cent of gross national product.
- Stress has a direct association with the autonomic system which controls a person's physiological and psychological responses to events – the 'flight or fight' syndrome.
- Certain occupational groups, such as teachers and nurses, are more prone to stress than other groups.
- Stress is very much concerned with how people adapt to changes in their lives, such as a new job, promotion, getting married, moving house or the death of a loved one.

2

The causes of stress

This chapter examines the causes of stress at work, the more common occupational stressors, sources of work stress and the problem of violence at work, commonly associated with bullying and, in some cases, having to deal with members of the public. The stress-related aspects of atypical working, including shift work and casual work, are also considered.

What is important to recognize is that individual stress responses vary considerably. No two people respond to the same stressor in the same way. These responses to stress are dealt with in Chapter 3. Current HSE recommendations relating to reducing stress at work are incorporated in this chapter.

2.1 Classification of the causes of stress at work

Stress affects people at work in many ways and the causes of stress are diverse. These causes can be associated with elements of the physical; environment, such as open plan office layouts, the way the organization is managed, relationships within the organization and even inadequate work equipment. The causes can be classified as follows.

2.1.1 The physical environment

Poor working conditions associated with the following can be frequent sources of stress in the workplace:

- Insufficient space to operate comfortably, safely and in the most efficient manner;
- Lack of privacy which may be disconcerting for some people;
- Open plan office layouts, resulting in distractions, noise, constant interruptions and difficulty in concentrating on the task in hand;

- Inhuman workplace layouts requiring excessive bending, stretching and manual handling of materials;
- Inadequate temperature and humidity control, creating excessive discomfort;
- Poor levels of illumination to the extent that tasks cannot be undertaken safely;
- Excessive noise levels, requiring the individual to raise his voice; and
- Inadequate ventilation, resulting in discomfort, particularly in summer months.

2.1.2 The organization

The organization, its policies and procedures, its culture and style of operation can be a cause of stress. Culture is defined as 'a state or set of manners in a particular organization'. All organizations incorporate one or more cultures, which may be described as, for example, friendly, hostile, unrewarding or family-style. Stress can be associated with organizational culture and style due to, for instance:

- Insufficient staff for the size of the workload, resulting in excessive overtime working;
- Too many unfilled posts, with employees having to 'double up' at tasks for which they have not necessarily been trained or instructed;
- Poor co-ordination between departments;
- Insufficient training to do the job well, creating uncertainty and lack of confidence in undertaking tasks;
- Inadequate information to the extent that people 'do not know where they stand';
- No control over the workload, the extent of which may fluctuate on a day-to-day basis;
- Rigid working procedures with no flexibility in approach; and
- No time being given to adjust to change, one of the greatest causes of stress amongst employees.

2.1.3 The way the organization is managed

Management styles, philosophies, work systems, approaches and objectives can contribute to the individual stress on employees, as a result of:

- Inconsistency in style and approach by different managers;
- Emphasis on competitiveness, often at the expense of safe and healthy working procedures;
- Crisis management all the time, due to management's inability, in many cases, to plan ahead and to manage sudden demands made by clients;
- Information being seen as power by some people, resulting in intentional withholding of key information which is relevant to tasks, procedures and systems;
- Procedures always being changed due, in many cases, to a failure by management to do the basic initial research into projects prior to commencement of same;
- Over-dependence on overtime working, on the presumption that employees are always amenable to the extra cash benefits to be derived from working overtime; and

- The need to operate shift work which can have a detrimental effect on the domestic lives of employees in some cases.

2.1.4 Role in the organization

Everyone has a role, function or purpose within the organization. Stress can be created through:

- Role ambiguity (see Chapter 3)
- Role conflict (see Chapter 3)
- Too little responsibility
- Lack of senior management support, particularly in the case of disciplinary matters dealt with by junior managers, such as supervisors, and
- Responsibility for people and things which some junior managers, in particular, may not have been adequately trained to deal with.

2.1.5 Relations within the organization

How people relate to each other within the organizational framework and structure can be a significant cause of stress, due to, perhaps:

- Poor relations with the boss which may arise through lack of understanding of each other's role and responsibilities, attitudes held, and other human emotions, such as greed, envy and lack of respect.
- Poor relations with colleagues and subordinates created by a wide range of human emotions.
- Difficulties in delegating responsibility due, perhaps, to lack of management training, the need 'to get the job done properly', lack of confidence in subordinates and no clear dividing lines as to the individual functions of management and employees.
- Personality conflicts arising from, for example, differences in language, regional accent, race, sex, temperament, level of education and knowledge.
- No feedback from colleagues or management, creating a feeling of isolation and despair.

2.1.6 Career development

Stress is directly related to progression or otherwise in a career within the organization. It may be created by:

- Lack of job security due to continuing changes within the organization's structure.
- Overpromotion due, perhaps, to incorrect selection or there being no one else available to fill the post effectively.
- Underpromotion, creating a feeling of 'having been overlooked'.

- Thwarted ambition, where the employee's personal ambitions do not necessarily tie up with management's perception of his current and future abilities.
- The job has insufficient status.
- Not being paid as well as others who do similar jobs.

2.1.7 Personal and social relationships

The relationships which exist between people on a personal and social basis are frequently a cause of stress through, for instance:

- Insufficient opportunities for social contact while at work due to the unremitting nature of tasks;
- Sexism and sexual harassment;
- Racism and racial harassment;
- Conflicts with family demands; and
- Divided loyalties between one's own needs and organizational demands.

2.1.8 Equipment

Inadequate, out-of-date, unreliable work equipment is frequently associated with stressful conditions amongst workers. Such equipment may be:

- Not suitable for the job or environment;
- Old and/or in poor condition;
- Unreliable or not properly maintained on a regular basis, resulting in constant breakdowns and down time;
- Badly sited, resulting in excessive manual handling of components or the need to walk excessive distances between different parts of a processing operation;
- Of such a design and sited in such a way that it requires the individual to adopt fixed and uncomfortable posture when operating same (see Chapter 7); and
- Adds to noise and heat levels, increasing discomfort and reducing effective verbal communication between employees.

2.1.9 Individual concerns

All people are different in terms of attitudes, personality, motivation and in their ability to cope with stressors. People may experience a stress response due to:

- Difficulty in coping with change;
- Lack of confidence in dealing with interpersonal problems, such as those arising from aggression, bullying and harassment at work;
- Not being assertive enough, allowing other people to dominate in terms of deciding how to do the work;

Table 2.1 The more common occupational stressors

New work patterns	Increased competition
New technology	Longer hours
Promotion	Redundancy
Relocation	Early retirement
Deregulation	Acquisition
Downsizing	Merger
Job design	Manning levels
Boredom	Insecurity
Noise	Lighting
Temperature	Atmosphere/ventilation

- Not being good at managing time, frequently resulting in pressure from supervisors and other employees to ensure the task is completed satisfactorily and on time; and
- Lack of knowledge about managing stress.

Some of the more common occupational stressors are shown in Table 2.1 above.

2.2 Factors contributing to stress at work

Not all stress is caused by work. In many cases, people bring their stress to work, which may be associated, for example, with:

- Financial problems, in particular, debt
- Single parenthood
- Relationship problems, such as marital separation or impending divorce
- Other family problems, such as children being in trouble with the police or their school, a sick parent or a child leaving home
- Moving house
- A death in the family
- Having a baby or infertility problems
- Serious or terminal illness
- Impending retirement
- Problems with getting child care.

In most cases, such persons are looking for a sympathetic response from their employer to their problems and many organizations now provide a range of support services, such as debt counselling, in addition to stress counselling for employees.

2.3 Categorizing the causes of stress

Occupational stress is more easily understood if a distinction is made between its cause and effects. Broadly, any situation at work which causes some form of stress response, and an increase in a person's level of arousal, is an occupational stressor.

Table 2.2 Classification of stressors

Type 1	Stressors that can be changed or eliminated with minimum effort, such as hunger, thirst, inadequate lighting or ventilation, excessive noise, members of a work group and badly fitting personal protective equipment
Type 2	Stressors that are difficult to change or eliminate, such as poor working relationships, financial problems, certain illnesses and conditions, inconsiderate managers and clients, technical difficulties with machinery and equipment and difficulties in separating work from home activities
Type 3	Stressors that are impossible to change, such as incurable illness, physical disabilities, death.

Some organizations use a system for categorizing stressors and the relative ease of eliminating or controlling individual causes of stress, as shown in Table 2.2 above.

The category or type of stress will determine the range and scale of support provided by the organization.

2.4 The main sources of work stress

Another way of categorizing stressors is on the basis of their source. Certain stressors impact on people through their senses, such as extremes of temperature, odours, noise, light and ventilation. Other stressors cause changes in thoughts and feelings, such as fear, excitement, arousal, ambiguity, threat and worry. A third group is associated with changes in body state, such as those created by illness, inputs of drugs, chemicals and alcohol.

Irrespective of the magnitude of each of these stressors, they create some form of impact and have a cumulative effect bringing the individual closer to his tolerance level for peak performance. Excessive input of stress takes the person beyond that peak tolerance level leading to some form of stress response.

The sources of stress vary considerably from person to person. However, a number of the more common sources of stress can be considered. These are:

- **Task-related factors**: work beyond the individual's mental capacity, information overload, boredom
- **Interpersonal factors**: day-to-day interaction with people, abuse and harassment
- **Role ambiguity**: the individual has no clear idea of what is expected of him (see Chapter 3)
- **Role conflict**: opposing demands made on an individual by different people
- **Little or no recognition** for a good job done
- **Personal threat**: actual threats to a person's safety, fear of redundancy or dismissal
- **Environmental factors**: noise, excessively high or low temperatures, inadequate lighting and ventilation, dirty workplaces, inadequate work space.

2.5 Recognizing stress in the workplace

What is the evidence of stress amongst employees that a manager should recognize? Observable behavioural symptoms include increases in medication and alcohol input and in smoking amongst certain people, habits such as nail biting, grinding of the teeth and chewing the inside of the mouth, bodily tics and fidgeting. Certain people may experience an increase in panic attacks. Others may demonstrate rapid body movements whilst walking. Above all, certain employees would appear to have lost their sense of humour, have become very 'touchy' and do not communicate as well with fellow employees as they did in the past.

An increase in the number of particularly minor accidents, such as slips, trips and falls, those associated with the use of hand tools and head injuries arising from inattention, is a fair indication of a workforce suffering stress.

2.6 Stress within the organization

Studies by Cooper and Marshall (1978) into sources of managerial stress identified an 'organizational boundary' with the individual manager straddling that boundary and, in effect, endeavouring to cope with conflicting stressors created by external demands (the family) and internal demands (the organization) (see Figure 2.1).

The manager's response may be affected by individual personality traits, his tolerance for ambiguity, his ability to cope with change, specific motivational factors and well-established behavioural patterns. Within the organization, a number of stressors can be present. These include those associated with:

1. The job
    - Too much or too little work
    - Poor physical working conditions
    - Time pressures
    - Decision-making, etc.
2. Role in the organization
    - Role conflict and role ambiguity
    - Responsibility for people
    - No participation in the decision-making, etc.
3. Career development
    - Overpromotion or underpromotion
    - Lack of job security
    - Thwarted ambition
4. Organizational structure and climate
    - Lack of effective consultation
    - Restrictions on behaviour
    - Office politics, etc.
5. Relations within the organization
    - poor relations with the boss
    - colleagues and subordinates
    - difficulties in delegating responsibility.

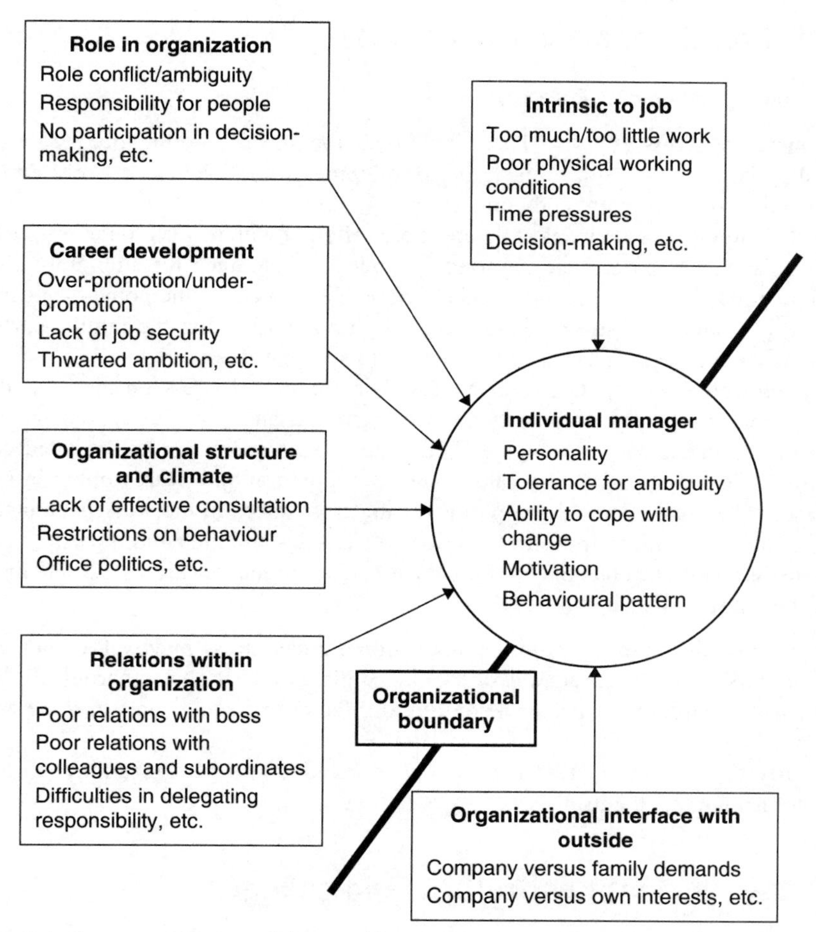

Figure 2.1 Sources of managerial stress (*Source*: Cooper and Marshall, 1978).

On the other side of the organizational boundary is the organization's interface with the outside world. Here conflict can be created where there may be competition for an individual's time between the organization and his family, or between the organization and an individual's own particular interests or hobbies.

The outcome of these studies is that organizations should pay attention to the potentially stressful effects of their decisions, management style, consultative arrangements, environmental levels and other matters which can have repercussions on people and their home lives. The resulting stress can have adverse effects on performance. In some cases, they may need to assist people in reconciling this 'home–work interface' through counselling and training in various coping strategies.

2.6.1 Transactional model of stress

This model (see Figure 2.2) depicts:

1. **Sources of stress at work**: These include factors intrinsic to the job, the individual's role in the organization, career development, relationships at work and the organizational structure and climate.
2. **Individual characteristics**: All people are different with respect to factors such as their levels of anxiety and neuroticism and their tolerance for ambiguity. Some people may demonstrate 'type A' behaviour. The effects of the home–work interface is a source of stress, characterized by problems within the family, partners endeavouring to balance careers and occasional life crises.
3. **Symptoms of occupational ill health**: The sources of stress on the individual, together with his individual's home–work circumstances and behaviour patterns, can result in excessive smoking and drinking, job dissatisfaction and reduced aspirations. This can lead to ill health, such as depression and heart trouble in some cases. The organizational symptoms arising from stress in the workplace include high labour turnover, industrial relations difficulties and high absenteeism.
4. **The diseases**: The outcome for individuals can be coronary heart disease and mental ill health (psychiatric injury).

In the case of the organization, a labour force under stress may well demonstrate their dissatisfaction by spasmodic or prolonged industrial action and chronically poor performance. Stress may also be a contributory factor in frequent and, in some cases, major accidents.

Clearly, the sources of stress at work must be tackled by the organization before the symptoms manifest themselves.

2.7 Organizational culture and change

All organizations incorporate a set of cultures that have developed over a period of time. They are associated with accepted standards of behaviour within the organization and, in many cases, are established by directors and senior managers. Considerable significance is attached by many organizations to the concept of defining, promoting and maintaining the right culture with respect to, for instance, quality, customer service, written communications and safety.

The term 'culture' can be defined in a number of ways:

- A state of manners, taste and intellectual development at a time or place (Collins Gem English Dictionary)
- Refinement or improvement of mind, tastes, etc. by education and training (Pocket Oxford Dictionary).

Culture is, however, not a static thing. It is continually changing with the emergence of new cultures brought about by new ideas, technologies and demands made on the organization in, for example, the market place, or as a result of new or modified legislation.

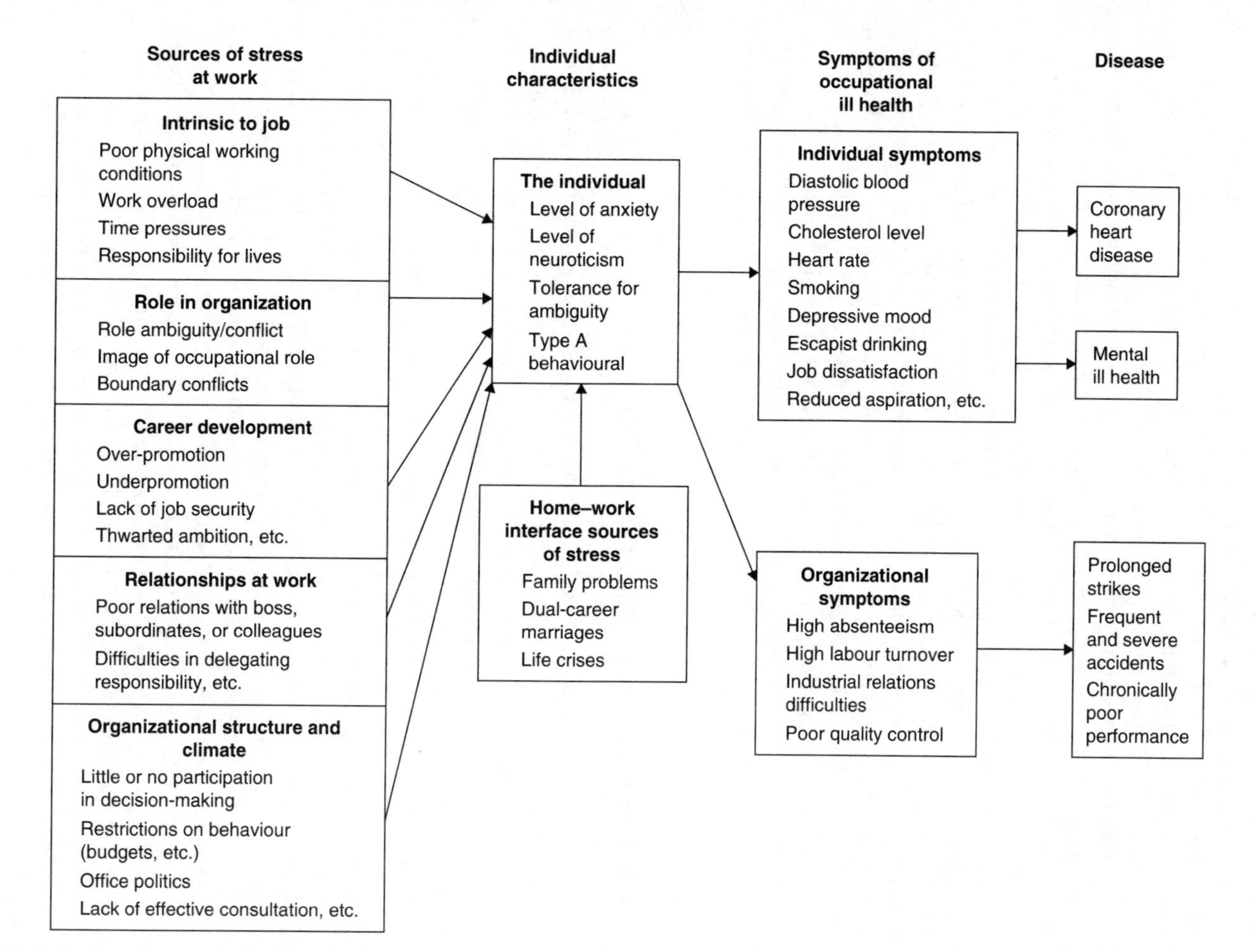

Figure 2.2 Transactional model of stress.

Culture is the result of all the daily conversations and interactions between members of an organization. People are continually agreeing or disagreeing about the 'proper' way to do things and how to make meaning of the events of the world around them. If employers wish to change the culture, then all these on-going conversations have to be changed or at least a substantial number of them. Out of these conversations arises a set of core beliefs which maintain the unity of the culture.

Organizations vary significantly in their cultures and management styles. Some are more progressive than others particularly when it comes to dealing with stress. A more positive culture will be concerned with the safety, health and welfare of its employees, viewing this area as an important feature of maintaining good employer–employee relationships, resulting in high levels of performance and productivity by employees. Managers recognize the significance of stress arising from change and make positive attempts to introduce change through consultation with the workforce on a regular basis. Regrettably, however, many organizations operate within a negative culture where management is highly resistant to change and employees feel helpless in any attempts to bring about change. This results in feelings of helplessness, cynicism towards management, stress amongst employees and poor standards of performance.

What is important to recognize is that organizational cultures cannot be changed according to plan, or through the demands of senior management, or by intervention by the enforcement authorities.

2.7.1 Bringing about cultural change

Many people have a prime function of bringing about change in an organization. These 'multi-purpose change agents' include a wide range of people, such as health and safety specialists, human resources managers, quality managers and trainers. Consultants are commonly brought into organizations with the principal objective of introducing change.

Generally, change can only be brought about very slowly. In many cases, it is a question of converting the 'hearts and minds' of employees to, for instance, a new system of working, the adoption of a particular safety practice or the use of new work equipment.

One of the principal causes of stress amongst employees is the attempt by the organization to enforce a new set of cultural norms and standards. This is frequently undertaken to a specific timetable, with predetermined criteria established with a view to measuring the success or otherwise in the various stages of the change process. Measurement may be undertaken through the use of employee questionnaires, observation and checking as to whether people are following the new system correctly and by obtaining feedback from line managers at regular meetings. This has been particularly the case in some organizations with the introduction of quality management systems, most of which are introduced on a staged basis. This is particularly stressful for older employees who may take a negative attitude to the change process and, in some cases, display hostility and cynicism.

A simple example of the introduction of change can be associated with the introduction of information technology in workplaces. Whilst the majority of the younger employees have mastered the use of computers, many older workers have found this exercise particularly stressful.

For cultural change to be successful, employers need to consider at the outset:

- The potentially stressful effects on the workforce of enforced change;
- The need for on-going consultation at the various stages of the change process;
- The provision of information, instruction and training prior to commencing the process;
- The speed at which the intended changes are scheduled to take place;
- Methods for assessing how well or how badly individual employees are coping with the changes;
- The provision of continuing help, assistance and coaching, together with the regular monitoring of people who are finding the changes difficult to comprehend and put into practice; and
- The provision of regular feedback to employees on the success or otherwise of the changes being introduced.

2.8 Stress in the work group

Many people work in well-established groups or teams. Evidence of Selye's 'flight or fight' response can frequently be identified in individual members of groups.

Groups with individuals who display a 'fight' response are characterized by aggressive behaviour, blaming and punishing activities, being excessively competitive with respect to other groups, operating a hidden agenda and being prone to industrial action.

Those groups with members who tend towards a 'flight' response may be seen to be acting in a withdrawn fashion, submissive to authority, uncommunicative and not contributing individually towards the success of the group. Members tend to ignore problems due to the lack of communication amongst members.

2.9 Shift workers and other atypical workers

'Atypical workers' are classified as those employees who are not in normal daytime employment, including shift workers, part-time workers and night workers. Approximately 29 per cent of employees work shifts and 25 per cent of employees undertake night shifts. Studies into the physical and psychological effects of atypical working amongst factory and transport workers has established that:

- Between 60 and 80 per cent of all shift workers experience longstanding sleep disorders;
- Shift workers are between five and 15 times more likely to experience mood disorders through poor quality sleep;
- Drug and alcohol abuse are higher with atypical workers;
- Eighty per cent complain of chronic fatigue;

- Approximately 75 per cent feel isolated from family and friends;
- Digestive disorders are four to five times more likely; and
- More serious accidents, resulting from human error, occur during shift work operations.

The psychological factors which affect an individual's ability to make the adjustments required to meet varying work schedules are associated with age, personal sleep needs, sex, the type of work and the extent of desynchronization of body rhythms, this last factor being the most significant.

2.9.1 Shift work

For the individual, the principal objective is to stabilize body rhythms and provide consistent time cues to the body. Employees should be trained to appreciate the stressful effects of shift working and that there is no perfect solution.

However, they do have some control over how they adjust their lives to the working arrangements and the changes in life style that this implies. They need to plan their sleeping, family and social contact schedules in such a way that the stress of this adjustment is minimized. Most adverse health effects arise as a result of changing daily schedules at a rate quicker than that at which the body can adjust. This can result in desynchronization with reduced efficiency due to sleep deprivation.

Important factors for consideration are:

- **Sleep deprivation**: This can have long-term effects on the health of the shift worker. The actual environment in which sleep takes place is important.
- **Diet**: A sensible dietary regime, taking account of the differences between the time of eating and the timing of the digestive system, will assist the worker to minimize digestive disorders.
- **Alcohol and drugs**: Avoidance of alcohol and drugs, that is caffeine and nicotine, results in improved sleep quality.
- **Family and friends**: Better planning of family and other social events is necessary to reduce the feeling of isolation frequently experienced by shift workers.

2.9.2 Part-time working

Part-time work is generally taken to mean work of less than 30 hours a week. Many people undertake part-time work of necessity to supplement their income or to provide a basic income. With the increase in single-parent families, many mothers of young children find it necessary to undertake part-time work in order to survive financially. For certain groups, part-time working means balancing the needs of the family with the need to generate extra income. This can be particularly stressful, particularly when rigidly set working periods create difficulties with child care arrangements or the need to look after elderly parents or relatives.

Moreover, part-time workers may be treated less favourably than full-time workers in terms of benefits, leave entitlements, promotion and training opportunities, giving them the feeling of being treated as 'second class citizens' by the organization.

A number of remedies are available to reduce the stress associated with part-time working. These include:

- **Home or teleworking**: regular working from home, or teleworking;
- **Term-time working**: allows an employee to work only during his children's school terms;
- **Voluntary reduced working time (or V-time)**: allows an individual to voluntarily agree to reduced working hours for a specific period, such as 12 months, with the right to return to full-time employment at the end of that period;
- **Compressed working hours**: where the working week is reduced to a defined number of hours;
- **Job sharing**: where the responsibilities of a particular job are shared by two people, for example one person working mornings and the other, afternoons; and
- **Annualized hours**: a system whereby the amount of time worked is defined over a whole year, the employee choosing, by agreement with the employer, when to actually undertake the work according to the needs of the business and his own convenience.

2.9.3 Night work

The physiological, psychological and medical effects of night work have been the subjects of numerous studies. It is generally agreed that, although the effects of night work vary considerably, depending upon the worker's age, economic situation and family commitments, regular night work principally causes abnormal fatigue and is liable to affect in many ways the health of the worker, whether male or female.

According to studies by the International Labour Organization (ILO), excessive fatigue arises from sleep disturbances and the fact that night workers have to work in a state of 'nocturnal deactivation' and sleep in a state of 'diurnal reactivation'. This provokes a discordance of phase between two circadian rhythms, the biological rhythm of the body's activation and deactivation and the artificial rhythm of activity at work and rest.

The severe sleepiness and tiredness experienced by night workers normally causes reduced alertness and consequently increases the risk of accidents. In rare cases, it may even cause 'night shift paralysis', an unusual phenomenon observed among air traffic controllers and night nurses, whereby the lack of sleep renders the individual unable to react to stimuli which would normally generate a reaction.

Undertaking night work has been shown to be related to digestive disorders, such as gastrointestinal conditions and, in particular, ulcers and nervous disorders which may be aggravated by an unsuitable diet, excessive consumption of coffee and excessive smoking during the night and by the use of sleeping pills during the day.

Other studies indicate an increased risk of cardiovascular diseases which is mainly attributed to the eating habits of shift workers and night workers.

The disturbance of family and social life adds to the psychological stress suffered by night workers with more or less serious consequences for their family relationships, life style and social adjustments. In many cases, night workers become isolated from the normal activities of their family. This arises through the fact that when things are happening within a family, like day-to-day activities such as eating together, watching television together or having discussions on a range of matters, the night worker is either at work or asleep. The only time the night worker will participate in family activities is at the weekend.

As a result of this isolation within the family, night workers can lose interest in family matters, their spouses and partners having to make decisions on important matters that normally they would make together. In many cases, the night worker is solely interested in his work and what goes on at work with no other interests inside or outside the family. This arrangement causes stress within other family members arising from this isolation of a member of the family.

As an ILO study on the subject concluded:

It appears to be well-established that, from both the physiological point of view and the family and social point of view, night work is harmful to the large majority of workers and is, therefore, to be deprecated.

2.9.4 Homeworking

The number of employees working at home, as opposed to a defined workplace, has increased significantly in recent years. It is likely to become more widespread in the light of the right of an employee to request flexible working introduced in the Employment Act 2002. In this case employers are required to give serious consideration to applications for flexible working (including homeworking) from qualifying employees who are parents of children under 6 years or disabled children under the age of 18. Where a request is refused an employer must be able to show he did so on objective business grounds in order to avoid any risk of legal action.

Whilst homeworking may seem an attractive proposition for many people, it is likely to be most successful if the particular employees are able to cope with the reduced social contact with work colleagues that this causes, are able to arrange family commitments to provide a suitable working environment, and are mature, self-motivated and self-disciplined.

The risks created by this form of social isolation must be taken into account before agreeing to homeworking. To counteract this, there must be excellent levels of communication between office and the homeworker with an agreed routine for regular meetings with immediate management, either at the permanent workplace or at the home of the homeworker.

Other factors which are relevant to homeworking, in terms of health and safety for instance, must be considered. This may require a risk assessment being undertaken by the employer, particularly where the homeworker may be using machinery, electrical equipment, hazardous substances, display screen equipment or, for instance, a room in the attic or an outbuilding to undertake the work.

2.10 The home–work interface

Achieving a happy balance between the demands of work and those of the home environment has, for many people, become difficult in the last decade. The introduction of E-mails, mobile phones and other forms of communication has aggravated the situation in that people are no longer able to divorce their home life from their work. People are being put under greater pressure at work and it is essential to achieve a reasonable balance between home and work.

According to Professor Cary Cooper of UMIST, stress arising from this home–work interface can result in:

- **Divided loyalties**: whereby employees are frequently required to make decisions in terms of their loyalties to the demands of the family as opposed to those arising from work.
- **Conflict of work with family demands**: particularly in the case of overtime working, resulting in employees spending more time at work instead of being with their families and participating in family activities, such as family outings and eating together.
- **Intrusion of problems outside work**: due to working excessive hours, the employee may not be in a position to deal with a range of matters which require his attention, including those of an economic nature or life crisis situations.

2.11 Reducing stress at organizational level

A number of remedies are available to organizations with respect to atypical working procedures. These include:

- Consultation prior to the introduction of shift work
- Recognition by management that shift work can be stressful for certain workers and groups of workers and of the need to assist their adjustment to this type of work
- Regular health surveillance to identify any health deterioration or changes at an early stage
- Training to recognize the potentially stressful effects and the changes in lifestyle that may be needed and
- Better communication between management and shift workers aimed at reducing the feeling of isolation.

Professor Cooper divides the organizational strategies for the management of workplace stress into three groups in order of importance. They are:

1. **Primary**: The first strategy is risk assessment or an organizational stress audit. It is aimed at eliminating or modifying environmental stressors to reduce their negative impact on individuals. The likely structural interventions would consist of:
   - Job redesign
   - Culture change

- Encouraging participative management
- Flexible working
- Work–life balance policies
- Organizational restructuring and
- Improved organizational communications.

2. **Secondary**: The secondary strategy is stress management training and health promotion. It focuses on increasing the awareness, resilience and coping skills of the individual through:
   - Stress management education and training so that the symptoms of stress can be recognized
   - Lifestyle information and health promotion activities
   - Skills training more generally, e.g. time management, presentation skills and
   - A reward-orientated management style.
3. **Tertiary**: The third strategy is workplace counselling and employee assistance programmes. It is concerned with the treatment and rehabilitation of distressed individuals, e.g. counselling and return to work policies.

According to Cooper, the strategies for dealing with stress at work can be summed up as follows:

1. Work should be more fun.
2. Employees are human beings who need praise. Therefore the fault-finding culture must be replaced by one that rewards.
3. Work–life balance.
4. Flexible work options.

2.12 Violence, bullying and harassment at work

2.12.1 Psychological and physical violence at work

Some organizations are characterized by an aggressive management culture which the majority of employees find stressful. Such a culture can lead to bullying and harassment, in effect, psychological violence. In some cases, employees may suffer both physical abuse, in terms of physical violence, and verbal abuse, from managers, other employees, customers and members of the public. Physical violence can result in horrific physical injuries. What people do not realize is, however, that psychological violence can, over a period of time, cause just as severe injury to health.

2.12.2 Current statistics

Although the risk of employees being physically assaulted at work is relatively low, approximately 1.3 million workplace assaults were recorded in England and Wales in 1999. According to the British Crime Survey, 2.5 per cent of employees were the victim of a violent act at work. Moreover, figures contained in a recently published joint survey from the Home Office and HSE (2002), *Violence at Work: New Findings from the 2000 British Crime Survey*, reveal that nearly 75 per cent of UK employees had received no formal training or even informal advice on how to safely deal with threatening behaviour at work. This survey further indicated that 17 per cent of employees who had some form of contact with members of the public were either very worried or fairly worried about being threatened and that the total number of violent workplace incidents has risen by 5 per cent between 1997 and 1999.

Although the latter fact is not statistically significant, it does suggest that the 19 per cent fall in violent incidents from 1995 to 1997 may have been reversed.

High risk professions and occupations are the police, social workers, probation officers, security guards and bar staff. Nurses and teachers face increasing violence. On average there are seven incidents of violence per month in each NHS Trust in England and Wales, adding up to around 65 000 incidents per year. Around two-thirds of attacks are on nurses.

2.12.3 Violent incidents at work each year

The latest statistics on violence at work were published in February 2004 in the British Crime Survey by the Home Office and HSE. Although the number of reported incidents of physical assault and threats has fallen since the peak in 1995, it remains at a worrying level.

Workers in the protective services, for example police officers, were most at risk of violence. Fourteen per cent of workers in protective services and 5 per cent of health- and welfare-associated professionals, including nurses, medical and dental practitioners, experienced violent behaviour.

Physical attacks are obviously dangerous, but serious or persistent verbal abuse can be a significant problem too as it can damage employees' health through anxiety and stress. For their employers this can represent a real financial cost through low staff morale and high staff turnover. This in turn can affect the confidence of a business and its profitability. Further costs may arise from increases in insurance premiums and compensation payments.

All work-related violence, both verbal and physical, has serious consequences for employers and for their business. For employees, violence can cause pain, distress and even disability or death.

2.12.4 Specific aspects of work-related violence

Evidence suggests that exposure to violence in the workplace is a significant contributor to work-related stress. This can arise in certain areas of work and in specific work situations.

Dealing with the public

The HSE publication *Violence at work: A Guide for Employers* defines work-related violence as 'any incident in which a person is abused, threatened or assaulted in circumstances relating to their work'. Verbal abuse and threats are the most common type of incident, whereas physical assaults are relatively rare.

People who deal directly with the public may face aggressive or violent behaviour. They may be sworn at, threatened or even attacked. Many employees, whose job requires them to deal with the public, are subject to some risk of violence, in particular:

- Those giving a service, such as working behind bars or counters
- People in caring occupations, such as nursing
- Teachers and others involved in education
- People involved in cash transactions, such as bank employees and those transferring cash, such as security guards
- Deliverers and collectors of goods to shops and other premises
- Staff involved in controlling members of the public, such as crowd attendants at football matches, railway platform staff and ticket inspectors and
- Those representing authority, for example, police officers and enforcement officers.

Bullying and harassment

Many managers would perceive the problem of bullying and harassment of their employees as primarily an industrial relations issue and, as such, should be dealt with through an employer's internal grievance and disciplinary procedures long before the problem becomes a risk to the health of those employees.

Levels of bullying vary significantly. Bullying is the common denominator of harassment, discrimination, prejudice, abuse, conflict and violence. It could be said that some people are 'serial bullies', whether they be managers or employees, and simply do not recognize this fact.

Employees must be in a position to report this form of behaviour confidentially to their employer with a view to seeking preventive action.

Stress caused by bullying can produce a number of symptoms in the victims:

- **Principal symptoms**: Stress, anxiety, sleeplessness, fatigue, including chronic fatigue syndrome (see later) and trauma.
- **Physical symptoms**: Aches and pains, with no clear cause, back pain, chest pains and angina, high blood pressure, headaches and migraines, sweating, palpitations, trembling and hormonal problems, etc.
- **Psychological symptoms**: Panic attacks, thoughts of suicide, stress breakdown, forgetfulness, impoverished or intermittently functioning memory, poor concentration, flashbacks and replays, excessive guilt, disbelief, confusion and bewilderment, insecurity, desperation, etc.
- **Behavioural symptoms**: Tearfulness, irritability, angry outbursts, obsessiveness (the experience takes over a person's life), hypervigilance, hypersensitivity to any remark made, sullenness, mood swings, withdrawal, indecision, loss of humour, etc.
- **Effects on personality**: Shattered self-confidence and self-esteem, low self-image, loss of self-worth and self-love.

Other symptoms and disorders include sleep disorder, mood disorder, eating disorder, anxiety disorder, panic disorder and skin disorder.

What is important is that the traumatizing effect of bullying results in the individual being unable to state clearly what is happening to him and who is responsible. The target of bullying may be so traumatized that he is unable to articulate his experiences for a year or more after the event. This often frustrates or prevents both disciplinary action and any subsequent legal action in respect of alleged post-traumatic stress disorder (PTSD).

Bullying commonly results in feelings of fear, shame, embarrassment and guilt which are encouraged by the bully in order to prevent his target raising the issue with management. In addition, work colleagues may often withdraw their support and then join in with the bullying, which increases the stress and consequent psychiatric injury. In addition, the prospect of going to work, or the thought or sound of the bully approaching, immediately activates the stress response, but flight or fight are both inappropriate. In cases of repeated bullying, the stress response prepares the body to respond physically, but is of little avail in most cases.

Management, therefore, have a duty to do something about this problem. The starting point is a policy on bullying and harassment at work which is brought to the attention of all employees. Managers should know what motivates the bully and take disciplinary action. Those suffering bullying would benefit from assertiveness training to defend themselves against unwarranted verbal and physical harassment.

2.12.5 Fatigue

People subject to bullying at work commonly suffer fatigue. This is caused by the body's flight or fight mechanism being activated for long periods, for example from Sunday evening, prior to starting work on Monday morning, right through to the following Saturday morning, when there is a chance to obtain some relief.

The flight or fight mechanism is, fundamentally, a brief and intermittent response. However, when activated for abnormally long periods, it can cause the body's mental, physical and emotional reserves to drain away. The body sustains damage through prolonged raised levels of glucocorticoids, which are toxic to brain cells, and excessive depletion of energy reserves resulting in loss of strength and stamina, fatigue and muscle wastage.

2.12.6 Chronic fatigue syndrome

People who are subject to bullying and harassment commonly suffer symptoms similar to chronic fatigue syndrome. (This is sometimes referred to as myalgic encephalomyelitis (ME), chronic fatigue immune deficiency syndrome (CFIDS) and post viral fatigue syndrome.)

This syndrome is characterized by:

- Overwhelming fatigue
- Joint and muscle pains
- Spasmodic bursts of energy, followed by exhaustion and joint and muscle pain

- A general lack of the ability to concentrate for long periods
- Poor recall with respect to words and sentence construction, for example
- Mood swings, including anger and depression
- Difficulty in taking on new information
- Imbalances in the senses of smell, appetite and taste
- A dislike of bright lights and excessive noise
- An inability to control body temperature
- Sleep disturbances, manifested by spending the night awake and then sleeping during the day
- Disturbances in the sense of balance
- Clumsiness, e.g. inability to grasp small objects.

2.12.7 Psychiatric injury

The symptoms described above eventually lead to psychiatric injury, but not mental illness. In fact, it may be appropriate at this stage to distinguish between psychiatric injury and mental illness. In spite of superficial similarity, there are distinct differences between psychiatric injury and mental illness. These include:

- Mental illness is assumed to be inherent (internal), whereas psychiatric injury is caused by external factors, such as bullying and harassment.
- Injuries tend to heal or get better.
- A person who is suffering a mental illness can exhibit a range of symptoms which are commonly associated with mental illness, such as schizophrenia, delusions and paranoia, but not those associated with psychiatric injury.
- A person suffering psychiatric injury, on the other hand, typically exhibits a range of symptoms, including obsessiveness, hypervigilance, irritability, fatigue, hypersensitivity and insomnia, commonly associated with psychiatric injury but not mental illness.

2.12.8 Suicide

People who suffer bullying have many common characteristics. For instance, they are unwilling to resort to violence, or even to resort to private legal action, to resolve conflict. They have a tendency to internalize anger rather than express it outwardly which, in many cases, leads to depression. Where the bullying continues over a long period of time, say several years, the internalized anger accumulates to the point where the victim may:

- Start to demonstrate all the symptoms of stress as the internal pressure causes the body to go out of stasis (this happens in every case); or
- Focuses the anger on to himself and harms himself, through the use of drugs and/or alcohol, or attempts to, or actually commits, suicide; or
- In rare cases, actually 'flip' and start to copy the behaviour patterns of the bully.

The lead up to suicide takes place in a series of steps. Firstly, bullying causes prolonged negative stress (psychiatric injury) which includes reactive depression. This results in a fluctuating baseline of the person's objectivity, that is, the balance of the mind is disturbed. At this stage, the victim starts to contemplate suicide culminating with the later stage where the victim actually attempts suicide. This is, of course, a cry for help and if this situation is not recognized and dealt with, actual suicide may follow.

It is conceivable that many suicides are caused by bullying. There is evidence that at least 16 children in the UK commit suicide as a result of bullying at school. In many cases, a victim of bullying is not initially aware of what is actually happening. He may be unwilling to confide with a colleague or manager what is happening to him. He may well become traumatized and unable to articulate. In many cases, the organization will deny the existence of bullying by individuals, or a culture of bullying from the top downwards, resulting in the real causes not being identified.

2.13 Violence management

The effective management of violence should take place in a series of stages.

2.13.1 Identifying the problem

All levels of management should be aware of what is going on in the workplaces for which they are responsible. They may consider that violence is not a problem or that incidents are uncommon. However, it is important to have a strategy for dealing with violence of all types and the following measures should be taken:

Discussions should take place on an informal basis with line managers and safety representatives. Employees may feel reluctant to discuss this issue so the use of a short questionnaire may be appropriate to establish whether employees ever feel threatened, firstly, by other employees and, secondly, by clients, customers and members of the public. It is important that the results of the survey are brought to their attention, indicating the recognition of the problem by their employer.

As part of induction training, and at regular periods afterwards, employees must be advised of the need to report all violent incidents to their manager.

These incidents of violent behaviour, including verbal abuse and threats, should be recorded on a specific form which enables the employee to give an account of the particular incident, the people involved, the location of the incident and any adverse outcomes, such as physical injury, fear, anxiety and stress-related symptoms, such as insomnia.

Regular analysis of reported incidents of violence should enable the employer to identify the work situations and locations where there is a potential for violent behaviour and, in many cases, the people responsible. This is particularly appropriate where employees are involved in serving customers, as in licensed premises, shops and offices, which may require the banning of certain customers from the premises and/or a formal complaint to the police.

Employers need to maintain statistical information on the outcome of these incidents, such as the scale of physical injury sustained, the number of days sickness absence resulting from the incident, emotional shock suffered, whether counselling was required and the period that the victim suffered distress or other forms of stress response. Regular analysis of records, assessment of reported incidents, discussions with employers running similar types of operation and public organizations, such as the local Chamber of Commerce, should enable the employer to establish a picture of the overall problem and enable him to decide on future action.

2.13.2 Risk assessment

The risk assessment process should take the following form:

1. It is important to identify those employees who are at risk. This could include those who have to deal directly with the public, such as across a counter or in face-to-face contact, or who have to visit clients in their own premises, such as the employees of estate agents.
2. It may be possible to measure the risk of violence on the basis of the nature of the work undertaken, such as advising clients or dealing with people making complaints about a range of matters. In particular, certain employees working on their own, such as anyone involved in collecting money from premises, may be particularly at risk.
3. In evaluating the risk of violence, it is necessary to consider existing protection arrangements. The views of the employees concerned should be sought as to whether the precautions already in place are adequate or whether more protection is required. This protection may take the form of extra training, improving the environment by, for example, upgrading illuminance levels in certain areas, physical security measures or modifications to the tasks undertaken.
4. Certain employees may benefit from assertiveness training to enable them to cope more effectively with adverse situations. All employees dealing with the public should be trained to identify the early signs of aggression and measures to avoid it or cope with it. They should be advised of clients with a past record of aggressive behaviour.
5. Public waiting areas, in particular, should be well-illuminated and maintained at a comfortable temperature.
6. It may be necessary to consider the installation of a number of security measures, such as closed-circuit television systems, personal alarms, coded locks on certain doors and deeper counters.
7. Where there may be an increased risk of physical violence arising from, for example, an attempted robbery, or where employees may work away from base, tasks should be modified, including varying the times at which these tasks are undertaken, ensuring people who transfer money to a bank are accompanied by another colleague, requiring employees away from base to report in on a frequent basis, providing transport home for people working late at night and eliminating as far as possible lone worker situations.
8. A record of the significant findings of the assessment must be maintained.

The risk assessment should be reviewed on a regular basis, particularly at the introduction of new working practices and procedures. In certain cases, it may be necessary to add further protective measures, or modify existing measures, as a result of experience and following consultation with employees.

2.13.3 Action

The first stage is the preparation of a Statement of Policy on Stress at Work. This should incorporate a statement of intent by the organization to take measures to eliminate or reduce stress as far as possible, the organization and arrangements for implementing the policy and the individual responsibilities of all levels of the organization in this respect. It is common practice to incorporate this statement as an appendix to the organization's Statement of Health and Safety Policy.

The risk assessment should identify the preventive and protective measures necessary with implementation of these measures on a phased basis.

2.13.4 Review

The success or otherwise of the measures above should be reviewed on a regular basis in conjunction with employees or their representatives. It may be appropriate to form a small committee to undertake this review exercise. Records of incidents should be maintained and examined regularly.

One outcome of the review process may be the need to return to the risk assessment stage because, for example, new work processes and systems have been introduced, the management structure has changed, employees may be experiencing overload or the number of violent incidents has increased.

2.13.5 Post-incident strategies

In the event of a violent incident involving one or more employees, employers must respond quickly in order to prevent any long-term distress. Such strategies include the provision of counselling, together with debriefing, allowing the individual to talk through the experience after the incident. No two people necessarily respond in the same way to both physical and mental violence and may need a limited amount of time off to recover.

The Home Office leaflet *Victims of Crime* gives more useful advice if one of your employees suffers an injury, loss or damage from a crime, including how to apply for compensation. It should be available from libraries, police stations, Citizens Advice Bureaux and victim support schemes. Further help may be available from victim support schemes that operate in many areas.

2.14 Conclusion

The causes of stress are many and varied. No two people respond to the same stressor in the same way. In the workplace, a host of factors may contribute to employee stress.

When undertaking risk assessments, employers need to consider the stress-related hazards to their employees and instigate strategies to prevent stress arising in the first place.

Questions to ask yourself after reading this chapter

- Have you considered the potential causes of stress to employees whilst at work?
- Do you have a good working relationship with each of your employees?
- Do you recognize that factors outside work, such as family conflict, can cause stress?
- What efforts does the organization make to recognize stress amongst employees?
- Is the organization going through a period of change? If so, what provision has been made to reduce the stressful effects of change on employees at all levels?
- Would you say the culture in your organization could be classed as 'friendly'?
- Do you pay particular attention to the specific needs and problems of shift workers, part-time workers and night workers?
- Are employees regularly contacted over work issues when they are at home?
- Has the organization ever been subject to a stress audit?
- Is there a problem with bullying and harassment at work? Do you have policies and procedures for dealing with this issue?

Key points – implications for employers

- The causes of stress can be associated with a range of factors – the physical environment, the organization, the way the organization is managed, an individual's role in the organization, relations within the organization, career development, personal and social relationships, work equipment and personal concerns.
- Shift workers and other atypical workers frequently suffer stress-related symptoms, such as chronic fatigue and poor quality sleep.
- Violence, bullying and harassment are a common feature of many organizations which employers have singly failed to recognize.
- Employees who have contact with members of the public frequently encounter violent behaviour in their work.
- Employers need to have formal procedures for dealing with work-related violence of all types.

3

Responses to stress

As stated earlier, no two people respond to the same stressor in the same way or with the same intensity of response. Moreover, the short-term response to stress by an individual may bear no relationship to his long-term response.

The long-term effects of stress on an organization must also be considered. Evidence of stress is common in many organizations, characterized by absenteeism, distrust of management, variable levels of production, high levels of sickness absence and labour turnover.

3.1 Symptoms of stress

Stress can have both short- and long-term responses. Individuals and, indeed, their employers should take the short-term symptoms seriously, bearing in mind the long-term implications, not only for individuals but for the organization as a whole. These symptoms are summarized in Table 3.1 below.

3.2 Responses to prolonged stress

As with any form of stressor, not only must the nature of the stressor be considered, but also the duration of exposure to the stressor. Physical responses to prolonged stress can include a number of minor disorders which create discomfort, but which may lead to serious ill health including headaches, migraine, allergies, skin disorders and arthritis.

The following diseases have been linked to stress, but there is no clear-cut medical evidence to this effect.

- **Coronary heart disease**: Studies have identified a positive link with competitive and aggressive behaviour and coronary heart disease.

Table 3.1 Short- and long-term responses to stress

Physiological and social	Mental, emotional, behavioural	Individual health
(a) Short-term response		
Headaches, migraine	Job dissatisfaction	
Backaches	Anxiety	
Eye and vision problems	Depression	
Allergic skin responses	Irritability	
Disturbed sleep patterns	Frustration	
Digestive disorders	Breakdown of relationships at home and work	
Raised heart rate	Alcohol and drug misuse	
Raised blood cholesterol	Tobacco smoking	
Raised adrenalin/ noradrenalin levels	Inability to unwind/relax	
(b) Long-term responses		
		Gastric/peptic ulcers
		Diabetes
		Arthritis
		Stroke
		High blood pressure
		Coronary heart disease
		Mental ill health

- **Cancer**: Some people, who are prone to symptoms of stress, such as anger, fear and feelings of hopelessness, may be more susceptible to cancer.
- **Digestive disorders**: Persistent indigestion or stomach discomfort is a classic manifestation of stress for some people. Approximately one person in 10 suffers from a stomach ulcer at some time in their life.
- **Diabetes**: Diabetes commonly follows some form of emotional or physical upset to the system.

The lessons to be learnt are that people need to be aware, firstly, of their own personal stress response and, secondly, to take positive action, rather than ignoring the evidence and pushing themselves even further.

3.3 The stages of the stress response

The response to stress commonly takes place in a number of stages in which a number of symptoms may be present. These are summarized in Table 3.2.

Managers should be trained to recognize some of the more obvious symptoms above as part of any stress reduction programme within an organization.

3.4 Stress indicators

Stress indicators are the danger signals that alert people to the fact that they are subjecting themselves to stress and pressure. These indicators can be physical, emotional

Table 3.2 Stages of the stress response

Stage 1	Speeding up Talking quickly Walking fast (head leading) Eating and drinking faster Working at high speed and for long periods of time without tiring (at the time)
Stage 2	Irritability Dyspepsia and gastric symptoms Tension headache Migraine Insomnia, loss of energy Comfort seeking – alcohol, smoking Increased intake of food and drink
Stage 3	Cottonwool head Gastric ulceration Palpitations, chest pain, cardiac incident Depression and anxiety Tiredness, lack of energy Physical and mental breakdown

or mental and are characterized by exaggerated or abnormal reactions to situations compared with their normal reactions.

Physical examples include tightness in the chest, muscle tightening, headaches, skin disorders, increased physical irritations, such as asthma, fidgeting, increased alcohol intake, loss or gain of appetite and disturbed sleep patterns. It is important to recognize that no two people display the same stress indicators. Emotional examples include increased irritability, anger, anxiety, frustration, touchiness and guilt. Mental examples include forgetfulness, trouble with thinking clearly, difficulty in forming sentences verbally and in writing, obsessiveness with petty detail and an overactive brain.

Stress indicators can also be recognized in an organization. For example, in some cases, senior management commonly fail to recognize the problem of stress amongst managers and employees. In other cases, they may take the view that stress is a manifestation of weakness on the part of an individual manager. Statements such as 'If you can't stand the heat, get out of the kitchen' are not uncommon amongst senior managers when presented with manifestations of stress on the part of subordinates.

However, stress within the organization cannot be disregarded. Typical indications of an organization manifesting high stress levels amongst employees at all levels are:

- Absenteeism
- Poor timekeeping
- High labour turnover
- High sickness absence rates
- Low productivity
- Industrial unrest
- Resistance to change in working procedures

Failure to recognize, and act on, these indicators leads to diminished levels of performance, diminished profitability for the organization and, for many people, long-term ill health.

3.5 The effects of stress on job performance

- **Absenteeism**: Absenteeism, especially on Monday mornings, or in the taking of early/extended meal breaks is a typical manifestation of stress.
- **Accidents**: People suffering stress at work can rapidly become problem drinkers. Such people have three times the average number of accidents; many accidents incorporate stress-related indirect causes.
- **Erratic job performance**: Alternating between low and high productivity due, in some cases, to changes outside the control of the individual, is a common symptom of stress within an organization.
- **Loss of concentration**: Stressful events in people's lives commonly result in a lack of the ability to concentrate, whereby a person is easily distracted, or an inability to complete one task at a time.
- **Loss of short-term memory**: This leads to arguments about who said, did or decided what.
- **Mistakes**: Stress is a classic cause of errors of judgement, which can result in accidents, wastage, rejects. Such mistakes are frequently blamed on others.
- **Personal appearance**: Becoming abnormally untidy, perhaps smelling of alcohol, is a common manifestation of a stressful state.
- **Poor staff relations**: People going through a period of stress frequently become irritable and sensitive to criticism. This may be accompanied by 'Jekyll and Hyde' mood changes, all of which have a direct effect on staff relationships and home life.

3.6 Anxiety and depression

Anxiety and depression are the classic manifestations of stress.

3.6.1 Anxiety and panic disorders

Anxiety is defined as 'a state of tension coupled with apprehension, worry, guilt, insecurity and a constant need for reassurance'. It is accompanied by a number of psychosomatic symptoms, such as profuse perspiration, difficulty in breathing, gastric disturbances, rapid heartbeat, frequent urination, muscle tension or high blood pressure. Insomnia is a reliable indicator of a state of anxiety.

Anxiety and panic disorders arise in many different forms, varying in intensity from person to person. They may be associated with certain disruptive life events, individual genetic features and changes in neurological chemistry. Emotional and physical stress can also lead to anxiety-related disorders.

Generalized anxiety disorder

This disorder is defined as 'a period of uncontrolled worry, nervousness and anxiety for a period of six months or more'. It may initially arise as a result of a particular worry concerning, for example, relationships with people, finance or career prospects, or may present as a vague anxiety about virtually everything. It is commonly accompanied by irritability, with a number of physical symptoms, including muscle pain, gastrointestinal problems, trembling and insomnia.

Panic disorder

Some people suffer panic attacks on a regular basis and these are diagnosed as panic disorder. When experiencing a panic attack, the individual suffers intense anxiety, fear or panic. Physical symptoms include an increased heart rate, perspiration, chest pains, tingling sensations and the feeling of disaster, loss of control or even imminent death. Some people who suffer panic attacks develop obsessive compulsive disorder or agoraphobia in an attempt to control the attacks. There are three different forms of panic attack: An *unexpected panic attack* can strike at any time and without warning. A *situational panic attack* always occurs in a specific location or situation, such as driving at night or when entering a particular building. The particular location or circumstance always provides the cue for a panic attack. A *situationally predisposed panic attack* also occurs in response to a particular situation or location, but does not occur every time the cue is present. The reaction may be delayed for some time.

Phobias

Phobias are, fundamentally, a fear response on the part of the individual to certain situations and things. People can have phobias to virtually anything or situation, such as fear of spiders, fear of flying, fear of heights and fear of enclosed spaces. The fear reaction generated by the phobia is out of proportion to the risks actually presented by the situation, and the person is aware of this fact intellectually. However, this awareness does not prevent phobias from occurring, and many people go to great lengths to avoid encountering the subject of their particular phobia.

Post-traumatic stress disorder (PTSD)

This is one of the few anxiety disorders where a specific trigger or cause for the disorder can be identified. It can be caused by highly stressful situations, such as those arising from war and disasters, physical attack, sight of death and other death situations, such as the death of a loved one. People suffering PTSD often relive the event that triggered the disorder initially, either through 'flashbacks' or nightmares. Some people make considerable efforts to avoid situations, places or circumstances that remind them of the traumatic event. Many people suffering PTSD develop an emotional insensitivity or 'numbness' as the mind attempts to protect itself from this disorder.

Obsessive compulsive disorder

This disorder affects a small number of people. People experiencing the disorder suffer persistent recurring thoughts caused by fears or anxiety. These repeating thoughts can, in some cases, lead to ritualistic behaviour designed to fend off the anxiety.

This includes excessive hand washing to maintain personal hygiene and prevent illness, checking and rechecking the security of the house and only using certain brands of goods.

Separation anxiety

This is a normal aspect of the development of children. Between the ages of approximately 8 to 14 months, children become aware of their need for comfortable and safe surroundings, in particular, the closeness of their mother and father. Children suffering separation anxiety cry when their parents are about to leave them in the presence of, for example, a babysitter. They cling on to their parents displaying various manifestations of anxiety. This anxiety, in some cases, persists to the age of 4 years or even later in childhood. It may re-emerge later in childhood under stressful situations, such as marital breakdown or when being left for the first time at boarding school.

When this form of behaviour becomes persistent and excessive, a child or young person may be diagnosed as suffering from separation anxiety disorder. They worry about being separated from their parents or from being taken from their homes or getting lost. They have trouble falling asleep, even in their own beds, and have a worry that their parents may die or abandon them.

Social anxiety disorder

This is a social phobia where sufferers are irrationally concerned about being judged or ridiculed in social circumstances and situations. They feel extreme embarrassment and anxiety when exposed to the outside world, and may suffer heart palpitations, sweating and blushing.

3.6.2 Depression

Depression, on the other hand, is much more a mood, characterized by feelings of dejection and gloom, and other permutations, such as feelings of hopelessness, futility and guilt. The well-known American psychiatrist, David Viscott, described depression as 'a sadness which has lost its relationship to the logical progression of events'. It may be mild or severe. Its milder form may be a direct result of a crisis in work relationships. Severe forms may exhibit biochemical disturbances, and the extreme form may lead to suicide. It is not a single disorder, however. There are different types of depression each manifesting a wide range of symptoms, each with varying degrees of severity. Symptoms of depression include sadness, anger, feelings of 'emptiness', pessimism about the future, low energy levels and sex drive and various forms of mental impairment, such as memory loss and difficulty in concentrating.

Depression may develop when one or more of the following factors are present:

- Family history of depression
- Personal history of depression
- Chronic stress
- Death of a loved one

- Chronic pain or illness or
- Drug or alcohol misuse.

Major depression

Where several of the above symptoms are present for a period of, for example, 2 weeks or more, an individual may be diagnosed by a psychiatrist or psychologist as suffering from major depression. This form of depression may occur perhaps once in a lifetime or, in some cases, may occur throughout life. Various combinations of therapy and medication are used to treat this condition.

Dysthymia

Symptoms are similar to those of a major depression, but tend to be less severe. The condition generally lasts longer than major depression, diagnosis of such being based on the presence of symptoms of depression in excess of 2 years. As with a major depression, dysthymia is treated with medication, therapy or a combination of both. Dysthymia is more resistant to treatment than major depression and can last longer if left untreated.

Bipolar disorder

This is commonly known as manic depression. Bipolar disorder is characterized by periods of depression interspersed with periods of mania. While depressive periods have similar symptoms to other types of depression, mania symptoms are quite different and include euphoria, a false sense of well-being, poor judgement, an unrealistic view of personal abilities and inappropriate social behaviour.

Post partum depression

Approximately 80 per cent of women experience the 'baby blues' after giving birth. Hormonal changes, the physical trauma of giving birth and the emotional strain of new responsibilities often make the first 3 or 4 weeks after a birth emotionally stressful. The situation eventually resolves without any need for treatment. For a small percentage of mothers, however, the baby blues do not go away, developing into post partum depression. If not successfully treated, this can lead to major depression or dysthymia.

Post partum psychosis

This condition is a rare complication of post partum depression, affecting a very small number of women after childbirth. The patient will display psychotic behaviour, often directed at the new baby. Patients may suffer hallucinations and delusions and the severity of the condition can sometimes put both mother and baby at risk.

3.6.3 Advice to managers

Managers frequently regard depression as being of no importance. Depression is not a sign of weakness, something which is untreatable, a personal failing or just feeling sad. It is a specific condition which needs sympathetic consideration, help and a responsible approach where employees report this condition.

3.7 Role theory

Role theory sees large organizations as systems of interlocking roles. These roles relate to what people do and what other people expect of them rather than their individual identities. An individual's thoughts and actions are influenced by identification with a role and with the duties and rights associated with that role. Everyone in a role has contact with people – superiors, subordinates, external contacts or contractors, who communicate their expectations of the role holder, trying to influence his behaviour and subjecting him to feedback. The individual, therefore, has certain expectations about how people should behave according to their status, age, function and responsibility. These expectations form the basis for a standard by which individual behaviour is evaluated, as well as a guide for reward. Stress arises in this framework due to role ambiguity, role conflict and role overload.

3.7.1 Role ambiguity

This is the situation where the role holder has insufficient information for adequate performance of his role, or where the information is open to more than one interpretation. Potentially ambiguous situations are in jobs where there is a time lag between the action taken and visible results, or where the role holder is unable to see the results of his actions.

3.7.2 Role conflict

Role conflict arises where members of the organization, who exchange information with the role holder, have different expectations of his role. Each may exert pressure on the role holder. Satisfying one expectation could make compliance with other expectations difficult. This is the classic 'servant of two masters' situation.

3.7.3 Role overload and role underload

Role overload results from a combination of role ambiguity and role conflict. The role holder works harder to clarify normal expectations or to satisfy conflicting priorities which are impossible to achieve within the time limits specified. Similarly, certain persons, who may have had a demanding job, may be shunted into a job where there is too much time available to complete an identified workload, resulting in boredom, excessive attention to minute details, as far as subordinates are concerned, and a general feeling of isolation. Role underload can be a significant cause of stress.

Research has shown that where experience of role ambiguity, conflict and overload/underload is high, then job satisfaction is low. This may well be coupled with anxiety and depression, factors which may add to the onset of stress-related conditions, such as peptic ulcers, coronary heart disease and nervous breakdowns.

3.8 Personality and stress

Personality is defined as 'the dynamic organization within the individual of the psychophysical systems that determine his characteristic behaviour and thought' (Allport, 1961).

3.8.1 Personality traits and their relationship to stress

Various types and traits of personality have been established by psychological researchers over the last 30 years. These are classified as shown in Table 3.3.

Research indicates that most people combine traits of more than one of these 'types' and so the definitions above can only be used as a guide. The type most at risk to stress is Type A and the characteristics can be summarized as follows:

- Excessive competitiveness and striving towards advancement and achievement
- Accentuating various key words in ordinary speech without real need and tending to utter the last few words of a sentence more rapidly than the opening part
- Continual drive towards ill-defined goals
- Preoccupation with deadlines for all sorts of task
- Intolerance of delays and postponements in arrangements
- A level of mental alertness which can easily progress to aggressive behaviour
- Permanent impatience with people and situations and
- Feelings of guilt when having a rest or relaxing.

Table 3.3 Classification of personality types

Type A: Ambitious	Active and energetic, impatient if he has to wait in a queue, conscientious, maintains high standards, time is a problem – there is never enough, often intolerant of others who may be slower
Type B: Placid	Quiet, very little worries them, often uncompetitive, put their worries into things they can alter and leave others to worry about the rest
Type C: Worrying	Nervous, highly strung, not very confident of self-ability, anxious about the future and of being able to cope
Type D: Carefree	Love variety, often athletic and daring, very little worries them, not concerned about the future
Type E: Suspicious	Dedicated and serious, very concerned with others' opinions of them, do not take criticisms kindly and remember such criticisms for a long time, distrust most people
Type F: Dependent	Bored with their own company, sensitive to surroundings, rely on others a great deal, people who interest them most are oddly unreliable, they find the people that they really need to be boring, do not respond easily to change
Type G: Fussy	Punctilious, conscientious and like a set routine, do not like change, any new problems throws them because there are no rules to follow, conventional and predictable, collect stamps and coins and keep them in beautifully ordered state, great believers in authority

	Tick the appropriate column						
	High		→				Low
	6	5	4	3	2	1	0
Do you always feel rushed?							
Do you walk fast?							
Are you competitive?							
Are you impatient?							
Do you go all out?							
Do you have no, or few, interests, outside work?							
Do you hide your feelings?							
Do you talk fast?							
Do you try to do more than one thing at a time?							
Are you hard driving?							
Do you anticipate others, e.g. interrupt, finish their entences?							
Are you emphatic in speech?							
Do you want recognition from others?							
Do you eat quickly?							
Are you ambitious?							
Do you look ahead to the next task?							

Figure 3.1 Type A behaviour questionnaire.

The potential for stress-related ill health, such as heart disease, is common among Type As, together with emotional distress, depression and anxiety.

One way of assessing whether an individual is manifesting Type A behaviour is through the use of a confidential questionnaire shown in Figure 3.1. It is important that the purpose of the questionnaire is explained to the person completing it and that it should be completed honestly. The outcome of the questionnaire should not be disclosed without the agreement of the person completing it.

3.9 Submission, assertion and aggression

A study by Bond and Kilty entitled *Practical Methods of Dealing with Stress* considered three features of personality, namely the 'submissive', 'assertive' and 'aggressive' features, and the types of behaviour commonly associated with people manifesting these features. A number of aspects of verbal and non-verbal communication were used to compare these features of personality as follows.

3.9.1 Content of speech

- **Aggressive**: Demanding, blaming, making threats, firmly stating own point of view as the right one; attacking, giving orders when it is not appropriate; deciding for others and they know it; being pushy, trying to force others to do things.
- **Submissive**: Repeating 'I'm sorry' and 'I'm afraid'; waffling and avoiding the point; backing down frequently; putting themselves down; complaining 'behind the scenes'; not saying what they want; going along with others to keep the peace or to be liked; agreeing to do things they don't want to do without negotiating (and doing them resentfully, badly, late or not at all).
- **Assertive**: Honest, open and to the point, saying 'No' when they want to; giving praise and criticism; sharing and taking responsibility for their own feelings; giving and accepting valid praise and constructive criticism; reflecting inappropriate feedback; stating what they want clearly, gently and firmly; acknowledging their own right (and that of others) to state what they want; standing up for themselves and those dependent on them.

3.9.2 Eye contact

- **Aggressive**: Glaring, staring, hard gaze; looking down from a height
- **Submissive**: Avoiding eye contact; looking up from a lower position
- **Assertive**: Gentle, direct, relaxed gaze; being at the same eye level whenever possible.

3.9.3 Posture

- **Aggressive**: Solid stance, perhaps hands on hips, feet firmly apart; moving uncomfortably close to emphasize points; trying to get physically higher, standing when the other is sitting
- **Submissive**: Round-shouldered, head down, chest cramped, slumped; staying at a lower level, for example, sitting when the other person is standing
- **Assertive**: Relaxed, upright, well-balanced, facing the other person directly at a distance acceptable to the other's cultural background.

3.9.4 Gestures

- **Aggressive**: Pointing, waving, poking with finger; clenched fist; sharp flicks of the wrist; hand-crunching handshakes; overhard jocular slaps on the back
- **Submissive**: Nervous fiddling; generally hands and arms turned in on self
- **Assertive**: Balanced, open, relaxed gestures to emphasize points.

3.9.5 Facial expression

- **Aggressive**: Tense, clenched teeth, frown; superior, indignant or angry expression
- **Submissive**: Nervous smile; apologetic; hang-dog or blank look
- **Assertive**: Relaxed, open, firm and pleasant.

3.9.6 Timing

- **Aggressive**: Interrupting, leaving no time for others to have their say; incessant chatter while bulldozing
- **Submissive**: Hesitating, leaving many gaps where others can butt in; waffling for a long time or keeping quiet
- **Assertive**: Concisely putting own point of view and allowing others to have their say.

3.9.7 Voice tone, volume, etc.

- **Aggressive**: Loud, sharp, firm and threatening
- **Submissive**: Quiet, strained, childlike
- **Assertive**: Low pitched, relaxed, firm, medium volume and gentle.

Clearly, those persons coming within the 'submissive' personality classification are more likely to suffer bullying and harassment at work resulting in stress-related conditions. As such, they would benefit from assertiveness training, in particular. On the other hand, people with 'aggressive' personality traits may need this fact drawing to their attention and, in some cases, some form of disciplinary action may be necessary. Management must, furthermore, promote the message that aggressive behaviour in the workplace is unacceptable, incorporating this into policies on violence at work.

3.10 Crisis

Crisis can be defined as 'a situation when something happens that requires major decisions to be made quickly'. No two people necessarily respond to a perceived crisis in the same way, however. Similarly, a crisis situation to one person may be something relatively minor to another.

3.10.1 Crisis situations

Typical crisis situations could be associated with sudden bad health, a client going bankrupt, loss of money, moving house at very short notice, family difficulties, being made redundant, or losing important documents, such as a cheque book, diary or credit cards.

Ideally, we need to identify crises before they arise and plan contingencies to cope, for instance having colleagues to stand in and cover the job in the event of unexpected illness, having alternative sources of finance available, retaining extra copies of important documents. Most crisis situations are, however, totally unexpected.

3.10.2 The process of transition (Wei-Chi)

Wei-Chi is the Chinese for crisis. The more literal translation is 'From danger comes opportunity', indicating a slow process of transition, the various stages of which can be stressful. These various stages are shown in Figure 3.2.

Crisis management has much to offer both the individual and the organization. It is important that people should understand and recognize these various stages with a view to designing strategies and systems to prevent their occurrence.

3.11 Alcohol misuse

Stress can result in increased alcohol consumption by employees. Not only does excessive alcohol consumption represent a serious risk to safety, but there is always the risk of becoming dependent on alcohol. Alcoholism is a true addiction and the alcoholic should be encouraged to obtain medical help and advice. Alcohol abuse leads to broken homes, broken marriages, lost jobs, a certain amount of crime and unhappiness generally for those who may come into contact with the alcoholic, together with varying degrees of physical and mental disease. On the other hand, it is a fact that many people consume very large quantities of alcohol throughout a long life without showing any apparent ill effects whatever and that, in most cases, alcoholism is a symptom rather than a disease.

The general, although by no means universally accepted, belief today is that the physical diseases brought about by the excessive consumption of alcohol are the result of its indirect effect in producing malnutrition rather than its direct toxic one. The repeated consumption of strong spirits, especially on an empty stomach, can lead to chronic gastritis, and possible inflammation of the intestines which interferes with the absorption of food substances, notably those in the vitamin B group. This, in turn, damages the nerve cells causing alcoholic neuritis, injury to the brain cells leading to certain forms of insanity and, in some cases, cirrhosis of the liver.

The alcoholic is not necessarily the person who becomes obviously drunk on frequent occasions, but is more commonly the person who drinks steadily throughout the day, often without any immediate effect being apparent to others. Later, however, symptoms which are partly due to physical effects, partly to the underlying neurosis which is at the root of the trouble in most cases, and partly social, begin to show themselves. The individual eats less and drinks more, often begins the day with vomiting or nausea which necessitates taking the first drink before he can face the public, his appearance tends to become bloated and the eyes are often red and congested.

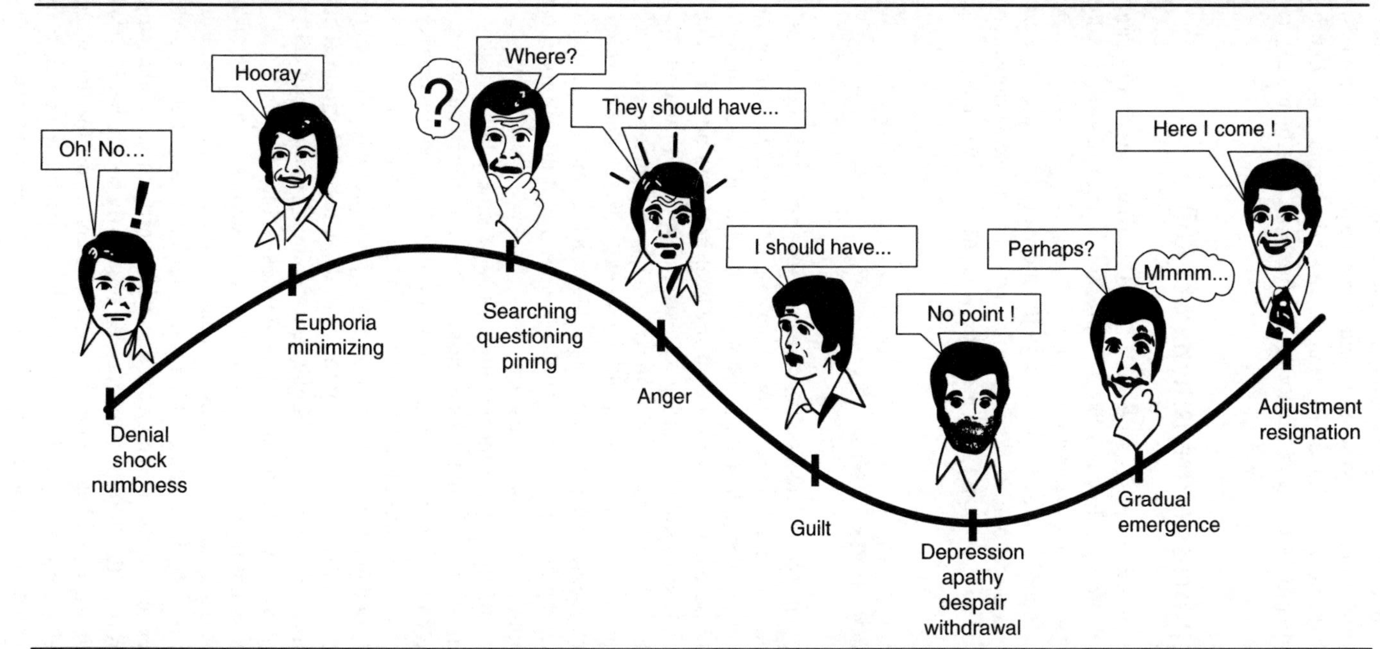

Figure 3.2 The process of transition.

His work suffers, he forgets to keep appointments and he becomes indifferent to his social responsibilities. His craving for drink becomes insatiable, and when he is unable to get it he becomes shaky, irritable and tense. Since he is ashamed of his condition, he tries to hide it and often, instead of drinking openly, hides the bottles about the house and, perhaps, his office. His emotions are less controlled and he gets angry or tearful readily, tells facile lies, and a minor illness or cessation of the supply may lead to an attack of delirium tremens (DTs).

In severe cases, the alcoholic may die from cirrhosis of the liver, or an attack of pneumonia or some other infection, not generally fatal to healthy people. No matter how alcoholism manifests itself, the alcoholic needs help, particularly if his condition is prejudicing the safety of his fellow workers. In most cases, this implies complete abstention for a period of time under controlled conditions, away from the normal temptations of the home and the workplace, perhaps psychotherapy to assess any psychological causes of the condition, and the general building up of impaired physical health.

It is in cases of alcoholism that the occupational health practitioner (occupational physician and occupational health nurse) can be of considerable support and assistance in bringing about the gradual rehabilitation necessary, perhaps through advising on the various social and therapeutic treatments available. The occupational health practitioner is also trained in the early detection of cases of alcohol abuse and, through counselling and routine surveillance, can prevent the situation from deteriorating further.

Many organizations have, over the last decade, produced statements of company policy on both smoking at work and alcohol at work, with a view to improving standards of employee health and reducing the sickness absence associated with smoking and alcohol consumption generally, and in the workplace and during working hours.

Figure 3.2 is a useful guide for managers where there may be evidence of alcohol addiction amongst staff. It shows the various phases in the development of alcohol addiction, the level of efficiency at these phases, various crisis points during the gradual deterioration of the individual and the visible signs during the various phases.

3.11.1 Alcohol at work

Alcohol misuse by employees, apart from its potential for creating accident situations at work, is a contributory factor in poor employee relations, poor discipline, absence and lateness amongst employees. Failure by employers to deal with excessive alcohol consumption, apart from the lost productivity and low performance it creates, can have adverse effects on morale and the organization's image in the market place. Organizations need, therefore, to implement a strategy for dealing with this problem which should take place in a series of steps.

Identifying the problem

In some organizations, it is not uncommon for employees to consume alcohol during actual working hours, before starting a shift, particularly a night shift, or even during meal breaks. Moreover, some employees live in a 'drinking culture', where they drink heavily and regularly with friends outside working hours, frequently getting drunk. They may return to work with a hangover. In many cases, the scale of the problem can be identified by reference to individual sickness absence records, whereby there may be a pattern of sickness absence, particularly on Mondays following heavy drinking sessions over the previous weekend. In some cases, certain employees may have suffered a greater number of minor accidents than the rest of the workforce, such as slips, trips and falls and contact accidents. Certain employees, such as buyers and sales representatives, may be exposed to alcohol consumption as part of their jobs. This may include entertaining clients, or being entertained by prospective suppliers.

Risk assessment

The risk assessment process should examine the hazards arising from the effects of alcohol consumption by employees with respect to their work activities, together with the risks to those employees and, indeed, to other employees, who may be affected by their drunken and unsafe actions. Moreover, certain tasks, such as those involving machinery operation or the driving of vehicles, require a greater attention to safety detail than others. Failure by employees under the influence of alcohol to take the appropriate precautions, or to indulge in careless behaviour, could have serious outcomes in terms of accidents and damage to plant, equipment and premises.

Action

Risk assessment should identify the preventive and protective measures necessary. In this case, employees with known drinking habits should be encouraged to receive help from an occupational health practitioner or specialist treatment organization. They may also need to be diverted to less demanding and safety-critical tasks until they have gained control over the habit. Employers must treat this matter with the same degree of confidentiality as they would any other health-related problem, and disciplinary action must be seen as the last resort.

It would be appropriate at this stage for an organization to produce a Statement of Policy on Alcohol at Work setting out the organization's approach to dealing with the problem including alcohol screening of employees. Line managers should be briefed on the rules and procedures outlined in the policy, the measures necessary where they suspect employees' drinking habits are affecting work and, most importantly, the consequences of their failure to take action.

Screening of employees at the pre-employment stage and afterwards should feature in the organization's occupational health procedures. This is a particularly sensitive issue and there must be the support and agreement from the workforce to its use as a protective measure before it is introduced.

Review

The frequent review of progress is essential to ensure the problem is being adequately controlled. Feedback from the workforce is essential here.

Model statement of policy on alcohol at work

This policy applies to all grades of staff and employees and to all types of work undertaken. Senior and line managers are responsible for implementing this policy.

This organization recognizes that the consumption of alcohol:

- during working hours
- during breaks
- when entertaining clients and
- on specific occasions at or arising from work

can have a detrimental effect on their work and represents a threat to both the health and safety of those employees and of others who may be affected by their actions whilst under the influence of alcohol.

All employees and other persons working on this site are required to take sensible precautions to protect themselves and other persons from hazards arising from the excessive consumption of alcohol. To this extent, the consumption of alcohol by any person is prohibited whilst at work and, in certain circumstances, employees who break this rule may be subject to disciplinary action.

The organization undertakes to provide professional help and assistance, in the strictest confidence, to any employee with an alcohol-related health condition. Further information on the hazards associated with alcohol is available from the occupational health nurse.

Date Signed

3.12 Drug misuse and addiction

The problem of drug misuse, commonly leading to addiction to drugs such as cocaine, morphine, heroin, etc., is common amongst certain age groups and ethnic groups. True addiction implies that the individual has developed a need for a particular drug in order to stay both physically and mentally sound. On removal of access to the particular drug, certain physical and/or psychiatric symptoms become apparent in the addict.

Addiction brings about changes in behaviour or bouts of abnormal behaviour and, whilst the individual needs careful attention from a doctor or drug specialist, employers must also take into account the risks to the safety of employees with whom the addict may come into contact whilst at work.

Initial health surveillance by an occupational health nurse will identify addiction situations.

The HSE (2001) publication *Drug Misuse at Work – A Guide for Employers* defines drug misuse as 'the *use* of illegal drugs and the *misuse*, whether deliberate or unintentional, of prescribed drugs and substances such as solvents'.

3.12.1 Signs of drug misuse

If employers are to tackle the problem of drug misuse amongst employees, it is important that they are aware of the signs. Fundamentally, drugs can affect both the brain and the body in a number of ways. For instance, they can alter the way an individual thinks, perceives and feels. This can lead to either impaired judgement or concentration. Misuse can result in the neglect of general health and well-being, adversely affecting performance at work, even when the misuse has actually taken place outside the workplace. Some of the common manifestations of the person misusing drugs are:

- An increased level of short-term sickness absence often accompanied by poor time-keeping;
- Occasional confused state;
- A marked deterioration in relationships with managers and colleagues at work;
- Sudden changes in mood and demeanour;
- Occasional aggressive behaviour;
- Irritability over minor events and situations;
- Reduced level of performance with varying levels of concentration with specific tasks;
- Energy level fluctuations.

In some cases, in order to maintain the habit, this can result in the individual committing a range of criminal offences, including theft of other people's money and goods and, in certain cases, physical assault with a view to robbery of individuals. However, it is important to recognize that all the above signs may not be caused by drug misuse. Many of these signs are classic manifestations of stress which may indicate that the person needs help.

3.12.2 Policies on drug misuse

With the extensive misuse of drugs in the UK, all organizations can benefit from a specific written policy applying to all persons, including visitors and, for example, employees of contractors, with respect to drug misuse. As with a policy on alcohol at work, such a policy should form part of the organization's Statement of Health and Safety Policy, perhaps as a specific appendix to it. Such a policy states the organization's approach to this problem, so that there is no possibility of misunderstanding or ambiguity on the subject. The policy should be brought to the attention of all employees even if there is no firm evidence at a particular point in time as to drug misuse by employees and other persons. Putting the policy into practice should take the following stages.

Identifying the problem

Most managers should have a fair idea of the presence or otherwise of drugs on site and the employees who may be misusing drugs. This may be evidenced by changes and fluctuations in behaviour amongst certain employees, varying levels of productivity by such persons, above-average levels of sickness absence, a higher level of accidents and 'near misses' in certain cases, and evidence of discord between individual employees.

Risk assessment

The risk assessment process should identify the hazards in terms of who is likely to be harmed by drug-related behaviour and the extent of the harm. The risk should then be measured and evaluated in terms of the future action that may be required, such as restricting known misusers of drugs to certain low risk operations.

Action

This entails implementing the formal Statement of Policy on Drug Misuse. A number of measures are required in this case. The first step must be that of increasing the awareness of all employees to the risks to health and safety of drug misuse by, for example, a series of short courses, group sessions and seminars at which the organization's policy would be explained. It should be stressed that anyone coming forward with a view to seeking help will be treated with utmost confidentiality and discretion.

In cases where employees do admit to drug misuse, they should be restricted to low risk activities and not be permitted to potentially dangerous work, such as the operation of machinery, lifting equipment and electrical equipment, driving vehicles of all types, such as fork lift trucks, or working at heights.

Line managers need to be advised on the signs of drug misuse and the organization's approach to dealing with this problem, including encouraging employees with a drug problem to seek professional help. Moreover, all managers must be advised of the criminal and civil law implications of permitting an employee to supply drugs to other employees and for a misuser to continue working, despite the risks he may create for himself and others.

The health and safety consultation process should ensure that trade union-appointed safety representatives and representatives of employee safety are fully informed of the organization's stance on this matter and of the need for co-operation in resolving the problem.

The process of resolving the problem should be co-ordinated by an occupational health practitioner, such as an occupational physician or occupational health nurse, in conjunction with the human resources manager and employee representatives. In certain cases, misusers may need time away from work for treatment.

Review

The risk assessment should be reviewed on a regular basis on the basis of information obtained since taking action and the need for any changes to procedures for dealing with drug misuse.

Model policy on drug misuse

This organization recognizes that drug misuse by employees and other persons on site represents not only a health risk to those persons, but can prejudice good standards of safety. On this basis, the organization is not prepared to tolerate drug misuse of any form by employees and other persons who may work, from time to time, on this site.

In this policy, the term 'drug misuse' means the use of illegal drugs and the misuse, whether deliberate or unintentional, of prescribed drugs and substances such as solvents (Health and Safety Executive).

The organization will provide all employees with information and instruction on the hazardous effects of drugs on the health and safety of themselves and other persons who may be affected by the actions of drug misusers.

Senior and line managers are responsible for implementing this policy in their respective areas of control. The Human Resources Director has overall responsibility for implementing this policy throughout the organization, in conjunction with the occupational physician and/or occupational health nurses.

All employees and other persons working on this site must act in such a manner as to ensure that drug misuse does not expose themselves or other persons to risks to their health or safety.

Any person found to be supplying drugs or dealing in drugs will be reported forthwith to the police and will be subject to disciplinary action, which may include dismissal for gross misconduct.

Employees seeking help must consult the occupational physician or occupational health nurse and will be treated with confidentiality and discretion at all times. Under these arrangements, periods of absence for treatment and rehabilitation will be regarded as normal sickness.

In specific cases of alleged drug misuse, or where the employee may undertake work where a high level of safety is required, the organization may require the employee to submit to drug screening and testing under the supervision of an occupational physician or occupational health nurse. Prior to such screening and testing, the employee concerned will be advised of the legal issues involved.

The effectiveness of implementation of this policy will be monitored by the occupational physician and/or occupational health nurse in conjunction with the Human Resources Director.

Date Signed

3.13 Women at work

Research has shown that women can be subject to many stressors at work which are not suffered by their male counterparts. Whilst sexual harassment at work is a common cause of stress amongst women, other causes of stress include:

- Performance related pressures;
- Lower rates of pay;

- The problem of maintaining dependants at home;
- Lack of encouragement from superiors, including not being taken seriously;
- Discrimination in terms of advancement;
- Sex discrimination and prejudice;
- Pressure from dependants at home;
- Career-related dilemmas, including whether to start a family or whether to marry/live with someone;
- Lack of social support from colleagues;
- Lack of same-sex role models;
- Evidence of male colleagues being treated more favourably by management;
- Being single and labelled as an oddity; and
- Lack of domestic support at home.

Management should be aware of the various forms of stress to which women are exposed at work and take measures, including disciplinary measures, where evidence of, particularly, sexual harassment exists.

Stress questionnaire for women at work

	Yes/No
1. Do you feel undervalued at work?	
2. Do you feel your employers don't take your concerns seriously?	
3. Do you feel you have been overlooked with regard to promotion?	
4. Do you suffer sexual harassment from colleagues and clients?	
5. Do you feel male colleagues are treated more favourably by your employer?	
6. Are there certain male colleagues you intentionally avoid at work?	
7. If you are not in a relationship, married or otherwise, do colleagues, both male and female, look down on you?	
8. Are you under pressure to perform to a particular standard?	
9. Do you feel you are poorly paid compared with male colleagues doing the same type of work?	
10. Do you have children?	
11. Do you have dependant relatives at home?	
12. Does your family make excessive demands on you?	
13. Do you suffer conflict from your desire to start a family but need to work?	
14. Do you have problems with child care, particularly during school holidays?	
15. Do you get any help at home with housework?	
16. Do you need 'a stiff drink' when you get home from work?	
17. Do you suffer from insomnia?	
18. Do you find it difficult to separate your work activities from home life?	
19. Does your manager regularly contact you at home over work issues?	
20. Do you feel permanently tired?	

Question: How many 'Yes' answers did you get?

Over 15: You are seriously stressed at work and need to consider alternatives, such as finding a less demanding job.

10–14: You need to discuss your work activities with your manager with a view to resolving some of your stress-related difficulties.

5–9: There are a number of aspects of your working life which you need to take in hand personally.

1–4: You have got a reasonably happy and stress-free working life. Try to do something about the aspects of work which cause you stress. Talk to your manager about it.

3.14 Conclusion

Employers need to be aware of the job factors which create stress and the classic responses to stress on the part of their employees. In particular, both anxiety and depression can arise from decisions made by employers, or even their lack of decision-making in some cases.

What is important for the organization is the effects of stress on job performance, which may be accompanied by, and associated with, excessive alcohol consumption and, in some cases, the taking of drugs by employees under stress.

Stress at work is a topic which is increasing in importance. Employers need to consider both the civil and criminal implications of stress at work indicated in Chapters 8 and 9 when making decisions which may possibly expose employees to stress.

Questions to ask yourself after reading this chapter

- What are the body's responses to prolonged stress?
- What are 'stress indicators'?
- What are the common effects of stress on job performance?
- Why do people suffer anxiety and depression?
- What are 'personality traits' and why are 'Type A' personality traits significant?
- Why is it important to deal with the problem of alcohol at work and excessive alcohol consumption by employees?
- What are the stressors which particularly apply to women at work?
- Why are role ambiguity, role conflict and role overload of particular importance in stress prevention?
- What are the stages of the stress response?
- What are the typical features of aggressive behaviour?

Key points – implications for employers

- A stressor causes stress. No two people respond to the same stressor in the same way.
- Symptoms of stress can be of a short- and long-term nature.
- Responses to long-term stress include coronary heart disease and digestive disorders.
- Stress indicators, such as muscle tightening, are the danger signals that alert people to the fact that they are subjecting themselves to stress and pressure.
- Evidence of stress in the organization is characterized by absenteeism, poor time-keeping, high labour turnover and low productivity.
- Anxiety and depression are the classic manifestations of stress.
- Stress arises in the organizational framework as a result of role ambiguity, role conflict and role overload/underload amongst employees.
- Certain personality types, particularly the Type A, are more prone to stress.
- Submission, assertion and aggression are features of personality which may be manifested by people at work.
- Reliance on smoking, alcohol and drugs is common amongst people encountering high levels of stress.
- Women employees may be subject to a range of stressors at work not experienced by their male counterparts.

4

The evaluation of stress

Many people go through stressful periods in their lives, some of which may be associated with their work, which require some form of change. Typical examples are being promoted to a higher position, greater responsibility for financial decisions or the classic takeover situation, whereby ownership of an organization changes hands, with all the uncertainty and worry about the security of jobs.

Wherever possible, organizations and individuals need to be able to measure and evaluate potentially stressful events and circumstances with a view to reducing their impact.

4.1 The measurement and evaluation of stress

Major changes in people's lives, such as marital separation, changes in responsibility at work, job loss and even getting married, can be stressful.

Research by Drs Holmes and Rahe of the School of Medicine, University of Washington, USA, into the clinical effects of major life changes has identified the concept of the life change unit (LCU), a unit of individual stress measurement in terms of the impact of stress on health. Over a period of 20 years Holmes and Rahe were able to assign a numerical value to a range of life events, such as a son or daughter leaving home, change in residence or death of a family member, and rank them according to their magnitude and importance. Then they compared the LCU scores of some 5000 individuals with their respective medical histories. They concluded that those with a high rating on the life change index were more likely to contract illness.

The Schedule of Recent Events (see Table 4.1) based on the total number of LCUs experienced in a year, has since been applied to many groups, confirming the view that the higher the degree of life change within a period of time, the greater the risk of subsequent illness, regardless of whether the change is perceived as desirable or undesirable.

Table 4.1 The Schedule of Recent Events (the Holmes–Rahe scale of life change units)

Event	LCU	Event	LCU
Death of a spouse	100	Change in work responsibilities	30
Marital separation	65	Son/daughter leaving home	29
Death of close family member	63	Trouble with in-laws	29
Personal injury or illness	53	Outstanding personal achievement	29
Marriage	50	Wife beginning or stopping work	29
Loss of job	47	Revision of personal habits	24
Marital reconciliation	45	Trouble with business superior	23
Retirement	45	Change in work hours/conditions	20
Change in health of a family member	44	Change in residence	20
Wife's pregnancy	40	Change in schools	20
Sex difficulties	39	Change in recreation	19
Gain of a new family member	39	Change in social activities	18
Change in financial status	38	Taking out a small mortgage	17
Death of a close friend	37	Change in sleeping habits	16
Change to a different kind of work	36	Change in number of family get-togethers	15
Increase or decrease in arguments	35	Change in eating habits with spouse	15
Taking out a bigger mortgage	31	Vacation	13
Foreclosure of mortgage or loan	30	Minor violations of law	11

Below 60, the individual's life has been unusually free from stress recently; 60–80, the individual has had a normal amount of stress recently. This score is average for an ordinary lifestyle; 80–100, the stress in the individual's life is a little high, possibly because of one recent event; over 100, pressures and stresses are piling up and the individual is under serious stress. They need to examine ways to reduce stress; over 200, this is a very high score and the individual is under serious stress so much so that he could be at risk of developing stress-related ill health. Measures should be taken to reduce the stress in his life; over 300, this is a serious score and the individual is at risk of developing stress-related ill health. Urgent measures should be taken to reduce stress.

According to Holmes and Rahe, if an individual's LCUs total 150–199, he stands a mild chance of illness in the following year, and 200–299, a moderate risk. Over 300 LCUs puts him in the group very likely to suffer serious physical or emotional illness. Specific scoring details are shown in the footnote to Table 4.1. The lessons to be learnt from this theory is that people should try to regulate the changes in their lives, most of which are under their control, and endeavour to stagger their incidence and intensity. The table shows life change events and LCU ratings. The values are, of course, averages.

4.2 Stress levels in occupations

It has long been held that some occupations are more stressful than others. Research into the question of these varying stress levels amongst different occupations was undertaken by the University of Manchester Institute of Science and Technology (UMIST) in the last decade. On a 0–10 scale, the following were identified as the more stressful professions and jobs (Table 4.2).

Table 4.2 The more stressful professions and jobs

Advertising	7.3	Broadcasting	6.8
Journalism	7.5	Musician	6.3
Actor	7.2	Film production	6.5
Dentistry	7.3	Doctor	6.8
Nursing/midwifery	6.5	Pilot (civil aviation)	7.5
Police	7.7	Ambulance service	6.3
Social worker	6.0	Teacher	6.2
Mining	8.3	Construction/building	7.5

UMIST (1987).

4.3 Conclusion

This chapter has covered some of the ways of measuring and evaluating stress. Other means are available for this purpose, but the scales and questionnaire indicated are most commonly used for this purpose.

It must be appreciated that particularly the Holmes–Rahe Scale of Life Change Events questionnaire and the Type A behaviour questionnaire should be used only by people who have been trained in the appropriate psychological techniques for measuring stress.

Questions to ask yourself after reading this chapter

- Is it possible to predict stressful events in people's lives?
- What is an LCU?
- How can Type A behaviour be measured?
- Are some occupations perceived as being more stressful than others?
- What occupations are classed as being the most stressful?

Key points – implications for employers

- The outcome of the measurement and evaluation of stress is an important element in making decisions with respect to preventing stress-induced ill health.
- The Schedule of Recent Events is commonly used to measure changes in people's lives which could be stressful.
- Stress levels vary dramatically in different occupations.
- Type 'A' behaviour, which can be measured by use of a questionnaire, is a classic manifestation of a person suffering from stress.

5

Coping with stress

It has been stated in previous chapters that, if people are going to cope with the stress in their lives, they need, firstly, to recognize those aspects of their lives which create their own personal stress response. However, some people may not even be aware of their personal stress response, such as insomnia, digestive disorders, a high level of arousal or a loss of appetite. They may attribute the palpitations, increased perspiration, sexual problems and depression to something totally different. It is essential, therefore, that any programme directed at raising the awareness of employees to stress should cover these aspects.

5.1 Responding to stress

In recent years, much research has been undertaken into the general area of stress and the measures individuals can take to reduce the stress in their lives. Adams' (1980) study, outlined below, is a well-respected approach to dealing with personal stress.

There are a number of ways for managing the stress in one's life. We are all unique and what works well for one person may be completely ineffective for another. Here is a range of ideas for responding to stress, each of which has worked well for someone, somewhere.

1. Become more knowledgeable about stress
   - Understand the process and effects of stress
   - Identify your major sources of stress – situations, people, etc.
   - Anticipate stressful periods and plan for them
   - Develop a repertoire of successful stress management techniques and practise them

  - Learn to identify the opportunities for personal growth inherent in periods of stress
  - Find the level of stress that is best for you, remembering that both insufficient and excessive stress are potentially harmful.
2. Take a systematic approach to problem solving
  - Define your problem specifically; divide it into manageable components that can be dealt with easily
  - Gather sufficient information about the problem and put it into perspective
  - Discover why the problem exists for you
  - Review your experience with the present problem or similar ones
  - Develop and evaluate a set of alternative courses of action
  - Select a course of action and proceed with it.
3. Come to terms with your feelings
  - Differentiate between your thoughts and your feelings
  - Do not suppress your feelings; acknowledge them to yourself, and share them with others
  - Learn to be flexible and adaptive
  - Accept your feelings.
4. Develop effective behavioural skills
  - Don't use the word *can't* when you actually mean *won't*
  - When you have determined what needs to be done with your life, act on your decisions
  - Use free time productively
  - Be assertive
  - Manage conflicts openly and directly
  - Avoid blaming others for situations
  - Provide positive feedback to others
  - Learn to say 'No'
  - Deal with problems as soon as they appear; if you procrastinate, they may intensify!
  - Evaluate the reality of your expectations, avoiding both the grandiose and the catastrophic
  - Learn to let go of situations and take breaks.
5. Establish and maintain a strong support network
  - Ask for direct help and be receptive when it is offered
  - Develop empathy for others
  - Make an honest assessment of your needs for support and satisfaction with the support you currently receive
  - List six people with whom you would like to improve your relationship and, in each case, identify one action step you will take toward such improvement
  - Rid yourself of dead and damaging relationships
  - Maintain high-quality relations
  - Tell the members of your support network that you value the relationships shared with them.
6. Develop a lifestyle that will buffer against the effects of stress

- Regularly practise some form of vigorous stretching and/or recreational exercise
- Engage regularly in some form of systematic relaxation
- Use alcohol in moderation or not at all
- Do not use tobacco
- Obtain sufficient rest on a regular basis
- Eat a balanced diet
- Avoid caffeine
- Avoid foods high in sugar, salt, white flour, saturated fats and chemicals
- Plan your use of time both on a daily and long-term basis
- Seek out variety and change of pace
- Take total responsibility for your life
- Maintain an optimistic attitude
- Do not dwell on unimportant matters.

7. Concentrate on positive spiritual development
   - Adopt the attitude that no problem is too monumental to be solved
   - Engage regularly in prayer or meditation
   - Establish a sense of purpose and relaxation
   - Seek spiritual guidance
   - Learn to transcend stressful situations
   - Believe in yourself
   - Increase your awareness of the interdependence of all things in the universe.
8. Plan and execute successful lifestyle changes
   - Expect to succeed; approach projects one step at a time
   - Keep change projects small and manageable
   - Practise each change rigorously for 21 days; then decide whether to continue with it
   - Celebrate your success. Reward yourself.

5.2 Personal coping strategies

A number of techniques and treatments are available to enable people to cope better with the stress in their lives. These include:

5.2.1 Relaxation training

This incorporates training people in deep relaxation techniques, relaxation during daily activities and emergency anxiety control. Such techniques incorporate aspects such as breathing control, physical (muscular) tension control and mind calming, including meditation.

- Choose a quiet place where you will not be disturbed.
- Lie down comfortably and ensure you are warm enough because as you relax your body temperature will fall slightly.

- Close your eyes and take three deep breaths in and sigh. This relaxes the diaphragm and therefore your breathing.
- Mentally go through your body physically tightening and then relaxing each part. Feel for areas of tenseness and then feel them relax on your 'out' breaths.
- Ignore outside noise interruptions by thinking of a lovely colour or a beautiful place or the sound of water. Let your mind and body float. If stray thoughts occur, just let them pass through your mind. For some people, the use of relaxation tapes, incense or aromatic oils can all aid relaxation.
- To recover, gradually deepen your breathing, start moving your muscles gently and as you 'awake' very gently arouse yourself. Get up by rolling on to your side and sitting before standing to avoid dizziness.
- When you are fully awake, stretch and take three deep breaths. After a few minutes you will feel refreshed and really alert. Ideally follow the relaxation period with a walk in the fresh air.

5.2.2 Physical exercise

A combination of physical exercises, such as walking, cycling, swimming, dance and aerobic exercise, together with those which raise the pulse and breathing rates significantly, such as squash, running and badminton, provide an excellent programme for stress control.

5.2.3 Drugs

The taking of drugs would not normally be recommended as part of a personal stress management programme unless prescribed by a registered medical practitioner. There is a risk of dependency and the individual may be distracted from coping with the problem through a false sense of well-being. In addition, the individual's ability to cope may be impaired through use of the drug.

5.3 Change management and stress

One of the principal causes of stress at work is the introduction of change. Change may be required in the case of working practices, it may come about as a result of the introduction of new technology, mergers between organizations, the introduction of new products or the revision of a management structure. How well or badly people cope with these changes in their lives, some of which are outside their control, is a key aspect of stress-related ill health.

In any consideration of bringing about change it is necessary to consider the effects on the organization generally and, secondly, the specific effects on individuals at all levels within the organization.

5.3.1 Obstacles to change

Michael Green, a consultant in the field of change management, in his paper *Practical Approaches to Meeting Business Needs of the Future* (1982) specifies eight obstacles to change within organizations, namely:

Lack of vision and mission

If the organization has no vision, no sense of what it wants to create in this world, then it cannot hope to inspire or encourage its management or staff to embrace the changes that are required. Without a vision there is no direction. Without a mission, there is no purpose.

Lack of corporate values

'How do we want to act, consistently with our mission, along the path towards achieving our vision?' Every organization needs a set of guiding principles marking out the way it wants to work on a day-to-day basis. These may include integrity, spirit of partnership, cost control and innovation. Whatever they are, be sure that, without them, you cannot have a properly managed business. In fact, without a shared and appropriate set of guiding principles, you will have people doing different things in different ways, leading to lack of consistency and conflict.

Lack of alignment

Lack of alignment occurs when individual and group goals and objectives are not tied into the overall corporate purpose. People do not work towards a shared vision or common aim. The efficiency and effectiveness of the company suffers, with tremendous cost implications.

Lack of attunement

If there is a set of core values or guiding principles then people need to behave consistently with them, relying on them to ensure the right decisions and believing that they can rely on others to do the same. Alignment and attunement produce reliability, consistency, cohesion and integrity.

Lack of leadership ability

If managing is about coping with complexity, then leadership is about mastering change. As the rate of change increases there is an urgent need for a new generation of leaders who have the ability to set lofty and strategic visions, to align people to those visions and motivate and inspire them to reach them. No leader. No vision. Mediocrity.

Top team ineffectiveness

As more and more organizations move towards less hierarchical, more flexible working arrangements, the need for team working grows. As always in change management, how the top team performs and behaves is of the utmost practical and symbolic importance.

Individual ineffectiveness

As in other areas of business, so too with change management. There will be occasions when individuals will be shown to be ineffective in moving the organizational goal posts. The reasons may be various and diverse: attitude problems, skill deficiencies, interpersonal conflict, self-confidence deficiency. Any top team member displaying one of these characteristics is a real liability to the process. It is usually financially and symbolically appropriate to invest time and effort in turning the individual around and some form of mentoring approach should be sought.

Staff resistance to change

There is a whole body of literature on resistance to change and the ways to overcome it. Research findings on this question suggest that once the vision and values are in place, the appropriate ways to tackle the (inevitable) resistance to change become clear.

If any of these obstacles stand in the way of change, then staff are bound to become stressed. Whenever people are faced with a change in any areas of their lives, the stress arises when they feel they lack the abilities and skills to deal with the new situation. The time when change is perceived as exciting is the occasion when people feel confident enough to start dealing with the new arrangements and, under these circumstances, people may positively relish the opportunity.

5.3.2 Introducing change

The notions of alignment and attunement mentioned above need to be thought through quite carefully before any major change process is started. One of the best ways of achieving these factors is to simply put out a questionnaire about the proposed changes. Managers are often quite frightened of this approach because they fear the response that will be obtained. The other alternative is to allow staff to give their views freely, whether on paper or at a meeting, to a facilitator who undertakes to pass that feedback upwards in the organization, without telling anybody who said what.

Fundamentally, changes must be 'sold' to people as being exciting because, although excitement is a form of stress, it is positive stress, whereas fear is always accompanied by negative stress. Excitement looks forward to the future but stress makes people cling to the past.

5.4 Organizational change

Change management within organizations entails thoughtful planning, sensitive implementation and, above all, consultation with, and involvement of, the people affected by the changes. Where, on the other hand, changes are forced on people, a wide range of problems can arise. Change must be realistic, achievable and measurable.

Managers must ensure that people affected by the change agree with, or at least understand, the need for change, and have a chance to decide how the change will be managed, and to be involved in the planning and implementation of the change.

Face-to-face communication techniques should be used to handle sensitive issues involved in organizational change management. E-mail and written communications are an extremely weak form of conveying and developing understanding.

For organizational change that entails new actions, objectives and processes for a group or team of people, workshops should be used to achieve understanding, involvement, plans, measurable aims, actions and commitment. These principles should be applied to very difficult changes which may entail redundancies, closures and integrating merged or acquired organizations. Senior managers should not be perceived to be hiding behind middle managers and various forms of written communication. Generally, managers who fail to consult and involve their employees in managing bad news are perceived as weak and lacking in integrity. Where people are treated with humanity and respect, they will reciprocate.

The principles of organizational change management can be summarized thus:

- At all times there must be involvement and agreed support from people within the system
- There must be understanding by everyone as to where the organization is at present
- There must be understanding by everyone as to where the organization wants to be, when and why, and the measures necessary to achieve this aim
- Development towards the above aim should be planned in achievable measurable stages
- There must be appropriate communication, implementation and involvement if changes are to be managed successfully.

5.5 Personal change

Many people, at various stages of their lives, are confronted with change. This may arise through being made redundant, through changes in the organization's structure, a new manager or taking on a new organizational culture. Whatever has happened, the individual needs to change behaviour which entails some form of transition process.

John Fisher's (1999) model of personal change, the Transition Curve, is an excellent analysis of how individuals deal with personal change. This model is an extremely useful reference for people dealing with personal change and for managers and organizations helping employees deal with personal change. The Transition Curve follows a series of stages as follows:

- **Anxiety**: This arises from the awareness that events lie outside an individual's range of understanding or control. People are unable to adequately picture what the future holds and they do not have enough information to allow them to anticipate behaving in a different way within the new organization. They are unsure as to how to adequately construe acting in the new work and social situations.
- **Happiness**: The awareness that a person's viewpoint is recognized and shared by others is two-fold. At the basic level there is relief that things are going to change and not continue as before. Whether the past is perceived negatively or positively,

there is still a feeling of anticipation and possibly excitement at the prospect of improvement. On another level, there is satisfaction of knowing that some of a person's thoughts about the old system were correct and that something is going to be done about it. In this phase, people generally expect the best and anticipate a bright future, placing their own construct system on to the change and seeing themselves succeeding. One of the dangers in this phase is that of the inappropriate psychological contract. People may read more into the change and will get more from the change than is actually the case. The organization needs to manage this phase and ensure that unrealistic expectations are managed and redefined in the organization's terms without alienating the individual.

- **Fear**: The awareness of imminent incidental change in an individual's core behavioural system implies that people need to act in a different manner. This will have an impact on both their self-perception and on how others externally see them. However, in the main, they see little change in their normal interactions and believe they will be operating in much the same way, merely choosing a more appropriate, but new, action.
- **Threat**: The awareness of an imminent comprehensive change in a person's core behavioural structures brings about perception of a major lifestyle change, one that will radically alter their future choices and other people's perception of them. They are unsure as to how they will be able to act or react in what is, potentially, a new and alien environment, one where the 'old rules' no longer apply and where no new rules are yet established.
- **Guilt**: Once the individual begins exploring his self-perception, how he acted or reacted in the past and looking at alternative interpretations, he begins to redefine his sense of self. This generally involves identifying his core beliefs and how closely he has been meeting them. Recognition of the inappropriateness of his previous actions and the implications for him as a person can cause guilt as he realizes the impact of his behaviour.
- **Depression**: This phase is characterized by a general lack of motivation and confusion. Individuals are uncertain as to what the future holds and how they can fit into the future 'world'. Their representations are inappropriate and the resultant undermining of their core sense of self leaves them adrift with no sense of identity and no clear vision of how to operate.
- **Hostility**: At this stage there is continued effort to validate social predictions that have already proved to be a failure. The problem here is that individuals continue to operate processes that have repeatedly failed to achieve a successful outcome and are no longer part of the new process or are surplus to the new way of working. The new processes are ignored at best and actively undermined at worst.
- **Denial**: This stage is defined by a lack of acceptance of any change and denial that there will be any impact on the person. People keep acting as if the change has not happened using old practices and processes and ignoring evidence or information contrary to their belief systems.

It can be seen from the transition curve that it is important for individuals to understand the impact that the change will have on their own personal construct systems, and for them to be able to work through the implications for their self-perception. Any change, no matter how small, has the potential to impact on an individual and may generate conflict between existing values and beliefs and anticipated altered ones.

One danger for the individual, team and organization arises when a person persists in operating a set of practices that have been consistently shown to fail (or result in an undesirable consequence) in the past and that do not help extend and elaborate their world view. Another danger area is that of denial where people continue operating as they always have, denying that there is any change at all. Both of these can have a detrimental impact on an organization trying to change the culture and focus of its people.

5.6 Better time management

Many people suffer stress through their failure to manage their time available effectively. As a result, there are those last minute crises because a report has not been finished on time, travel arrangements have not been made for an impending business trip or too much time was spent doing a task which is enjoyable at the expense of another task considered a chore. In some cases, that which is urgent is confused with that which is important (but not urgent).

The outcome is panic, frustration because items cannot be found, feelings of hopelessness, loss of concentration and anger at how things have worked out. A number of remedies are available to improve time management.

- Prioritize and allocate time. Do not confuse the urgent with the important.
- Say 'No' more often than 'Yes' to people seeking to take up time.
- Concentrate on one thing at a time.
- Break major tasks down into small bits. Do one small bit at a time.
- Make a 'to do' list each day and try to stick to it. Delegate to others if possible.
- Manage other people's expectations, especially customers, to suit your time as well as theirs.
- Control your use of time rather than letting time control you.

5.7 Dealing with personal crisis

Many people go through periods of crisis, which may be due to a range of situations, some of which may be associated with poor time management, as indicated above, or as a result of unexpected events in their lives. People deal with crises in their lives in different ways, but the few simple points outlined below should help.

- Stop and think. Take several deep breaths. This helps to dispel the immediate physical reactions arising from crisis.
- Collect and check the facts. Evaluate their consequences. Are there any historical precedents to draw on?
- Identify your objectives. What are the most important objectives?

- Plan ahead. Identify what needs to be done and in what order.
- Put your plan into operation. Make sure it happens. This is the hardest bit.
- Communicate with the people concerned. Who else is involved and/or needs to know?
- Evaluate how well you handled the crisis. Modify your contingency plan for the future if necessary.

5.7.1 Things to do to solve stressful situations

There are a number of ways people can help solve stressful situations or the anticipation of these situations.

- Monitoring physical flexibility will reduce tension pains in the shoulders, back, neck and head. Start the day with 5 minutes of stretching exercises, especially in these areas of the body. This releases natural endorphins promoting a positive attitude and strengthening the immune system. If you spend most of your time at work sitting down, a short walk can help.
- Simple relaxation exercises lying on the floor, before going to bed, or whenever feeling stressed, can be of great help.
- Breathing exercises, again lying on the floor, and in a relaxed position, over a period of 5 minutes, increases the oxygen flow to the lungs and vital organs, reducing the physical symptoms of anxiety.
- Meditation techniques have physiological effects on the brain which combat the 'flight or fight' stress response. Sit comfortably with the eyes closed, taking deep breaths initially.

5.7.2 Things to do to improve time management

Failure to manage the time available is one of the principal causes of stress. How often do you hear people say 'If only I had time'. Why is it that some people are perpetually 'in a state' prior to meetings, before going on holiday or going to work? In many cases, they have simply failed to organize and plan their time effectively prior to these events which result in this stressful state. A number of strategies are available:

- **Allocate time for planning and organizing**: Taking time to think and plan ahead is time well spent. Do it now instead of putting it off till tomorrow.
- **Set goals**: Goals give direction to people's activities. They should be specific, realistic and achievable. Check to see whether you have achieved these goals.
- **Prioritize**: Use the 80–20 rule stated by the Italian economist Vilfredo Pareto who noted that 80 per cent of the reward comes from 20 per cent of the effort. Prioritize your time to concentrate on that 20 per cent which produces the best return.
- **Make a list**: Some people keep on top of their time by a simple daily 'to do' list which they assemble last thing on the previous day or first thing that morning. Sticking to the items on the list, avoiding distraction from it, is important.

- **Be flexible**: Allow time for interruptions and distractions. By virtue of the nature of the job, it may only be feasible to plan ahead for 50 per cent or less of one's time, taking into account potential interruptions from colleagues, telephones and the unplanned 'emergency'.
- **Consider your biological prime time**: Consider the time of day when you are at your best in terms of performance. Some people are morning people, others are night people. Knowing the best time of the day assists in planning future performance.
- **Do the right thing right**: Peter Drucker, the well-known management guru, says 'doing the right thing is more important than doing things right'. Doing the right thing is effectiveness; doing things right is efficiency. Focus first on effectiveness, then concentrate on efficiency.
- **Eliminate the urgent**: Urgent tasks have short-term consequences while important tasks are those with long-term goal-related implications. The objective must be to reduce those tasks classified as urgent to enable the more important priorities to be dealt with.
- **Practise the art of 'intelligent neglect'**: Efforts should be made to eliminate trivial tasks or those tasks which do not have long-term consequences. It may be possible to delegate some of these more trivial tasks and work on those tasks which require your specialized attention.
- **Avoid being a perfectionist**: People who set themselves standards of performance based on perfection by, for example, paying unnecessary attention to detail, inevitably suffer stress when these standards cannot be achieved. It may be necessary to reduce your expectations on a range of issues.
- **Conquer procrastination**: One technique to try is the 'Swiss cheese' method. When you are avoiding something, break it into smaller tasks or set a timer and work on the larger tasks for just 15 minutes. By doing a little at a time, eventually you will reach a point where you want to finish.
- **Learn to say 'No'**: Too many people suffer from overload due to a simple inability to say 'No' to further demands on their time. Learning to say 'No' to trivial, less important demands, assists in reducing the stress from overload.
- **Reward yourself**: Setting goals that are measurable and achievable is important. When these goals have been achieved, indulge in the reward promised.

5.8 Assertiveness training

Many people suffer feelings of inferiority, lack of confidence, self-doubt and lack of assertiveness. These feelings may have arisen in childhood, perhaps as a result of dominant parents, being bullied at school, racial discrimination at school or as a result of sibling rivalry. They commonly apologize when it is not their fault, frequently blame themselves and are not prepared to take charge of or remedy a situation.

Once this personality trait is established, as with any aspect of personality, it is very difficult to change, particularly in later life. For some people, assertiveness training may be the answer. However, it may not bring about permanent changes in the way people look at themselves and it is common for people to revert to their original submissive behaviour afterwards.

For assertiveness training to be effective, there must be regular reviews of progress in behaviour modification, including attempts to remove self-doubt, analysis of specific submissive behaviour, feedback on how well or otherwise the individual dealt with a particular situation and advice on improving self-confidence.

5.9 Coping strategies

A number of personal coping strategies have been covered previously. These, and other strategies, can be summarized as follows:

- When you feel stressed, practise taking long deep breaths.
- Take regular breaks from your work.
- Get regular exercise.
- Eat a balanced diet.
- Avoid caffeine, which is a stimulant.
- Avoid dependence on alcohol and drugs to help you relax. This can quickly become a crutch.
- Practise good time management and organizational skills.
- Use humour to lighten difficult situations.
- Seek to find the positive in every situation. View adversity as an opportunity for learning and growth.
- Do not bury or hide your emotions. Unresolved emotions can emerge as nightmares or physical illness.
- If you find that a relationship makes you stressed, end it if possible.
- Give compliments freely and smile often. This has a direct effect on mood.
- Learn to really listen to what people are saying rather than getting upset because you disagree with them. Seek to find areas of common ground and work towards a compromise.

5.10 Relaxation therapy

Many people prefer a method of relaxation to other treatments that may be costly, invasive or have side effects. Relaxation therapy has been used in many situations for the relief of chronic pain and insomnia and can include a range of techniques. Many people use this technique to reduce stress and anxiety. The methods may be deep or brief. Deep methods include autogenic training, meditation and progressive muscular relaxation (PMR). Brief methods include paced respiration and self-controlled relaxation.

The objective is to use the power of the mind and body to achieve a sense of relaxation. Relaxation therapies often focus on repeating a sound, word or prayer. They may focus on body sensation, lower the metabolism and make a person feel relaxed.

5.10.1 Deep relaxation method

This method of relaxation focuses on both the mind and the body. Some of the types of meditation include:

- **Autogenic training**, in which a person imagines being in a peaceful place with pleasant body sensations. The person focuses on the body and tries to make parts of the body feel heavy, warm or cool. Breathing is centred and the heart beat is regulated.
- **Mindfulness meditation**, in which a person concentrates on body sensations and thoughts that occur in the moment. The person learns to observe sensations and thoughts without judging them.
- **Yoga or walking meditation**, which both come from Zen Buddhism and use physical discipline to focus the body and mind. Controlled breathing and slow deliberate movements and postures are used.
- **Progressive muscular relaxation**, in which a person focuses on tensing and relaxing each of the major muscle groups.
- **Transcendental meditation**, in which a person focuses on a sound or thought. A word, mantra or sound is repeated many times.
- **Biofeedback**, in which an instrument is used to monitor certain changes in the body, such as skin temperature or brainwaves. The person uses this information to try to relax deeply.

5.10.2 Brief relaxation method

Brief meditation requires much less time and skill. It is often a shorter form of a deep method of relaxation and includes:

- **Self-control meditation**, a shortened form of progressive muscular relaxation.
- **Paced respiration**, in which the subject breathes slowly and deliberately.
- **Deep breathing**, in which the person takes a deep breath, holds it for 3–5 seconds, then slowly releases it. This sequence is repeated several times.

Brief methods are often used when the individual faces stress or anxiety.

5.11 Ideas for managing stress

Whilst time management and assertiveness training are of considerable help in enabling the individual to deal with stress at work, there are many other personal stress management tools that people can use.

- Don't let things dominate you, such as the need to get a report finished or to study statistical information.
- View life as a series of challenges to seek, not obstacles that need to be avoided.
- Engage in regular exercise that is convenient and which gives pleasure, such as swimming.
- Maintain regular contact with friends who support you in the event of problems at work.

- Maintain a reasonable diet and ensure you get plenty of sleep.
- Take responsibility for your life and your emotions, but never blame yourself.
- Protect your personal freedoms and space. Do what you want and feel is right, but respect the rights of others.
- Don't tell others what to do, but if they intrude, tell them so!
- Surround yourself with cues from positive thoughts and relaxation.
- Open yourself to new experiences. Try new places, new things, new food and drink and enjoy meeting new people.
- Take short periods of private relaxation at regular intervals during the day.
- Don't drift along in stressful relationships or situations. Do something about it!
- Review your obligations and commitments from time to time and make sure they are still good for you.
- Avoid the use of medication, such as sleeping pills, tranquillizers and other stress-relieving drugs wherever possible.
- Above all, when things aren't working out and stress is building up, TALK TO SOMEONE!

5.12 Stress: what you can do

Most people get help on an informal basis from spouses, family and friends when exposed to a stressful situation which may be work-related or otherwise. In some cases, however, they may feel the need to talk to someone outside this circle. Where an individual is experiencing some sort of stress response, such as insomnia, loss of appetite or a feeling of hopelessness, it is better not to wait for the problem to increase but to talk to someone straight away.

A number of options are available. If it is a problem associated with work or home circumstances, it should be discussed with the person's immediate manager or HR manager so that, at least, they are aware of what that person is going through. Such matters should, of course, be treated in confidence in line with an organization's stress policy statement.

Where an occupational health service is provided, it would be appropriate to discuss the matter with an occupational health nurse or occupational physician. Both these occupational health professionals would be trained in stress counselling and can be of immense support. This is particularly appropriate where an individual may be suffering bullying or harassment at work or where he has been exposed to a stressful event, such as witnessing a serious or fatal accident at work.

In some cases, a person's trade union may be able to provide immediate help through an external agency.

5.13 Conclusion

Everyone is subject to stress at some time in their lives. It may be associated with a range of incidents and situations, such as loss of a job, death of a loved one, problems at work, bullying and harassment by colleagues at work or simply overloading of the job.

People would be much happier if they studied the causes of stress in their lives and adapted appropriate strategies for coping with these stressors. This chapter has provided a number of strategies for coping, including personal relaxation and time management, two important aspects of reducing those occasional crises that arise.

Questions to ask yourself after reading this chapter

- What do people need to do in order to cope more successfully with the stress in their lives?
- What does relaxation therapy entail?
- Why can organizational change be a source of stress for everyone?
- What are the effects of change on people?
- How can time management be improved?
- What do people need to do in personal crisis situations?
- Do some people need assertiveness training?
- What forms of relaxation therapy are available?
- How do people cope with changes in their personal lives?
- How do people respond to stress?

Key points – implications for employers

- For people to cope with the stress in their lives, they must be aware of those factors which create stress (stressors) and their specific stress responses, such as insomnia, digestive disorders or even aggressive behaviour.

- In order to deal with stress, everyone should be trained to develop their own personal coping strategies.

- Change within the organization, in job content, in the management systems and in working arrangements are some of the greatest causes of stress amongst employees.

- Better time management is one of the best ways of coping with stress at work.

- Some people, particularly those who have to deal with members of the public, may benefit from assertiveness training.

6

Stress in the workplace

There is clear-cut evidence from a range of publications, articles in the national press, reported cases in the civil courts and the increasing attention given to stress by the media that many people experience work-related stress at a level they believe is making them ill. The costs to society are substantial.

6.1 Advice to employers

Employers need to recognize the fact that certain aspects of the work, management culture and style, the process of innovation and how the organization communicates with employees are significant in the creation of stress. There are a number of strategies that they need to consider, therefore, with a view to preventing or reducing workplace stress.

6.1.1 Management style

Management style varies dramatically from one organization to another. Bullying and harassment of employees in order to achieve established objectives is common. However, stress will be considerably less where organizations are perceived by the workforce as caring and considerate. This approach will, in turn, be reflected in high levels of productivity and commitment by the workforce to the organization's success. Each organization must establish its own solutions in line with its particular management style and working practices. In some cases, there may be a need for drastic change to an organization's style and management culture.

6.1.2 Approaches and attitudes to stress

Regrettably, some managers take the view that stress is a sign of weakness which will not be tolerated. On this basis, it is important that managers understand the problem, treat it seriously and give their commitment to addressing the problem. It is important to ensure that individuals are not made to feel guilty about their stress problems and are encouraged to seek relief and support.

6.1.3 The right job

Some jobs, such as those which entail dealing with members of the public and clients are, without doubt, more stressful than others. People need to know the principal elements of their job, be confident that they can do it effectively and be given credit for successfully undertaking allocated tasks.

Under the Management of Health and Safety at Work Regulations, employers are required to take into account the individual capabilities of employees as regards health and safety when allocating tasks. This means that employers must consider both the physical capability and mental capability of the person involved before allocating that task. This could mean giving the person scope to make changes and involving him in any changes that might be made. The requirements of the job and the experience of the person must be considered, together with any information, instruction and training required. Any particularly stressful elements of the job should be explained, together with the measures the employee should take to relieve these elements.

6.1.4 Organizing change

Uncertainty as to future employment prospects is a principal cause of stress amongst employees at all levels. This arises during periods of major organizational change which employees know, or suspect, will affect them significantly in terms of redundancy, relocation or reassignment to new jobs.

Effective communication is the answer to reducing the stress associated with change. Employees need to be kept informed of the changes taking place or which are planned for the future.

6.2 Strategies for reducing stress

Any strategy directed at reducing stress amongst the workforce should follow a set pattern. The first stage is the identification of those work factors which cause stress, such as the current management culture and style, inflexible work schedules, inadequate communication, both upwards and downwards in the organization, and incompetent managers.

Having identified the stressors, they should then be measured and evaluated. This should take place through the use of employee questionnaires, discussions with individual employees, review of working practices and sickness absence rates, and meetings at which employees are encouraged to talk about the aspects of the job which cause stress.

This stage of the exercise should be followed by the organization declaring its intention to tackle the problem, commencing with the development and publication of a formal Statement of Policy on Stress at Work. Practical implementation of the policy should take account of the following.

6.2.1 Management culture and style

Objectives should be established which are both measurable and achievable with respect to, for example, improving communication and the involvement of employees, particularly prior to and during periods of change within the organization. This should be supported by the provision of information and training.

6.2.2 Work schedules

There may be a need to examine current work schedules with a view to increasing their flexibility and setting down the agreed working periods. Targets for production should be reasonable.

6.2.3 Relationships between employees

Some employees lack interpersonal skills and, as a result, may be subject to bullying, sexual and racial harassment. They would benefit from training in improving their interpersonal skills.

It may also be necessary to introduce formal procedures for dealing with bullying and harassment through the organization's grievance procedure and to ensure there is an effective system for investigating complaints of this nature by employees.

6.2.4 Consultation and planning on jobs

The internal consultation process should ensure that employees are consulted in the design and organization of their jobs and where new jobs are being established.

6.2.5 Ergonomic considerations

Tasks should be clearly defined taking into account ergonomic principles and with a view to reducing physical and mental fatigue.

6.2.6 The employee's role

Well-written job descriptions are essential in order to define the aims of a particular job, the individual responsibilities of employees in undertaking tasks and the support and supervision that will be provided. In many cases, training may be necessary for specific groups of employees, for example, those in regular contact with members of the public.

Line managers also need to be confident about the support that will be provided by senior managers with respect to ensuring the care of the employees for whom they are responsible.

6.2.7 Providing support

Regular monitoring of work activities and discussions with individual employees should identify those persons affected by stress and the pressures of work. Many people may need support and should be encouraged to attend both stress awareness and personal stress management courses. Above all, managers need to be briefed on the availability of internal and external support services, such as occupational health services, including those offering counselling. In smaller companies, employees suffering stress-related symptoms should be encouraged to see their doctor who, in most cases, will be able to provide support in conjunction with local agencies.

6.3 Recognizing stress in the workplace

There are many manifestations of stress in a workplace. These include:

6.3.1 The behaviour of employees

Tensions between individual employees may become increasingly apparent. These tensions and conflicts need resolving before they reach the stage where some form of disciplinary action may be necessary.

Stress may further be manifested in the attitudes to work of certain employees who may display anxiety, aggression or evidence of a depressed state (see 'Human behaviour and stress' later in this chapter).

6.3.2 Poor levels of performance

Reductions in output or productivity are common in organizations where employees are under stress. This may be further evidenced by poor levels of housekeeping, increased error rates, excessive waste, lack of effective decision-making, loss of motivation and commitment in the case of some employees and poor time-keeping. Some individuals may display a 'couldn't care less' attitude to both the organization and the work they undertake.

6.3.3 Sickness absence

An increase in overall sickness absence, in particular people taking frequent periods of short-term sickness absence, is a classic manifestation of an organization under stress.

Monitoring these three aspects can indicate whether stress levels are reducing and whether stress levels are higher in some areas than others. Sickness absence monitoring should further indicate problem areas.

6.4 The need to consider human factors

According to the HSE, human factors refers to 'environmental, organizational and job factors, and human and individual characteristics which influence behaviour at work in a way which can affect health and safety'.

In any workplace, and during workplace activity, there are complex interactions between factors relating to job, individual and organization. This complex interaction can have profound health and safety implications.

Tasks need to be ergonomically designed and matched to the capabilities of the individual employees carrying them out. This matching of the job to the person should involve both a *physical match*, including the design of the workplace and the working environment, and a *mental match*, involving the individual's information and decision-making requirements and their perception of the task and its risks.

Individuals vary with regard to their attitudes, skills, habits and personalities and these differences can have important influences upon task-related behaviour. Sometimes these influences are straightforward and obvious. However, often they are complex and much more difficult to identify. Whereas some factors, such as skills and attitudes, can be influenced and modified, others, such as personality, are much more fixed and resistant.

Although organizational factors have a major influence over individual and group behaviour, they are often ignored. This is true both at the design stage and at the investigation stage following an accident or incident. Cultural factors can exert enormous pressures upon individuals and lead them to adopt behaviour patterns out of a wish to conform. Unfortunately, these cultural pressures can be both negative and positive with regard to health and safety matters. There is little point in a culture which urges conformity to the rules, if those rules are themselves suspect from a health and safety standpoint.

As work equipment design has improved and work procedures have become tighter, so the emphasis has shifted on to human factors. According to the HSE, 'it is estimated that 80 per cent of accidents may be attributed, at least in part, to the acts or omissions of people'. However, although people may be directly involved it is often the underlying organizational problems that are at the root of the matter. These organizational issues often involve design, managerial and decision-making failures. Trust between employer and employees and open, two-way communications are vital if organizational failures are to be identified and rectified before they lead to human failures and consequently to accidents.

The HSE believe that human failures are caused by human error and violations. Examples of accidents involving human failure include the Three Mile Island nuclear reactor incident, the King's Cross underground station fire, the Clapham Junction rail crash and the sinking of the *Herald of Free Enterprise*. Although a human failing was the direct cause of the accident in each case, underlying organizational failures led to the human failing.

Human failings can lead to active or latent consequences. Latent failures can, for example, occur at the design stage and not lead to serious problems until much later. Poor training can also, ultimately, lead to latent failure. Active failures have immediate consequences, such as failing to observe the 'no smoking' rule when working in a flammable atmosphere.

6.5 Human behaviour and stress

Everyone is different. No two people perceive situations in the same way. Personalities vary significantly as do memory. Individual behavioural factors, such as attitude, motivation, perception, personality and memory, are all significant in any consideration of stress at work and, in particular, the measures taken by people to respond to stressors.

6.5.1 Attitude

Attitudes are an important feature of human behaviour. Many definitions of the term 'attitude' have been put forward over the years, such as:

- A predetermined set of responses built up as a result of experience of similar situations.
- A tendency to respond in a given way in a particular situation.
- The disposition of people to view things in a certain way and act accordingly.
- A mental and neural state of readiness, organized through experience, exerting a direct or dynamic influence upon an individual's response to all objects and situations with which it is related.
- A learned orientation, or disposition, toward an object or situation which provides a tendency to respond favourably or unfavourably to the object or situation.

Attitude comprises:

- A *cognitive* component, which is concerned with thoughts and knowing, such as perceiving, remembering, imagining, conceiving, judging, reasoning, the analysis of problems and the decision-making process; and
- An *effective* component which is concerned with emotions or feelings of attraction or revulsion.

The functions of attitude

According to Katz, there are four functions of attitude:

1. **The social adjustive function**: This is concerned with how people relate and adjust to the influences of parents, teachers, friends, colleagues and their superiors. It is argued that, by the age of 9 years, most attitudes are established. Behaviour is

based, to some extent, on a philosophy of 'maximum reward, minimum punishment'. This function is important in terms of how people deal with stress which may be created by their superiors at work.
2. **The value expressive function**: People use their attitudes to present a picture of themselves that is satisfying and pleasing to them. This is an important feature of attitude in that people see themselves as better and different, in some special way, from others around them. This is an aspect of an individual's 'self-image'. To promote this self-image, people may adopt extreme views of situations, dress in a particular way and adopt a particular political persuasion.
3. **Knowledge function**: Attitudes are used to provide a system of standards that organize and stabilize a world of changing experiences. On this basis, people need to work within an acceptable framework, have a scale of values and generally know where they stand. Where any of these three aspects are unacceptable, uncertain or obscure, stress may arise.
4. **Self-defensive function**: This function is concerned with a person's need to defend his self-image, both externally, in terms of how people react towards him, and internally, to deal with inner impulses and his own personal knowledge of what he is really like. Again, adverse responses from other people, such as work colleagues, may produce stress if the individual feels his self-image is threatened.

Attitude formation, development and change

Attitudes are formed as a result of continuing experience of situations during a lifetime and, as such, are difficult to change. They are directly associated with:

- Self-image – the image that the individual wishes to project to the outside world, for example, well-mannered, hard to please, stern, 'cool' or affluent.
- The influence of groups and the group norms, or standards, upheld by a particular group, such as people working in a department, whereby membership of the group and social contact with its members entails sharing their attitudes and conforming with the norms or standards.
- Individual opinions – an opinion is defined as 'what one thinks about something' or 'an attitude towards something which is very hard to change'; people carry a wide range of opinions which may be subject to challenge, mockery or correction putting the individual under some degree of stress.
- Superstitions – described as 'an opinion or practice based on belief in luck or magic'. People have superstitions, such as not walking under a ladder or fear of two crossed knives, despite the fact that there is no logical explanation as to why they should fear such situations.

Attitudes are very difficult to change. In many cases, people simply do not wish to change an attitude to a particular situation despite overwhelming evidence to support such a change. To be successful, attitude change must take place in a series of well-controlled stages.

One of the barriers to attitude change is 'cognitive dissonance', the situation that results when a person holds an attitude that is incompatible with the information presented. The theory of cognitive dissonance was proposed by Leon Festinger who postulated that, when faced with two pieces of information, that is, 'cognitions' (knowledge, thoughts, feelings) that are inconsistent, opposite or conflicting, people experience discomfort or stress. They will, therefore, want to reduce this discomfort or 'dissonance' by engaging in a variety of activities, such as changing their ideas, beliefs, knowledge or skills, or by avoiding the thoughts or cognitions altogether. In this way, many people endeavour to reduce or eliminate stress in a range of situations.

A number of factors are important in the attitude change process.

The individual

Built-in opinion

People acquire opinions on a range of topics from their parents, at school and in the work situation which, over a period of time, become hardened. Attempting to change these opinions, which may be one of the objectives of a particular training course, for example, can be stressful both for the trainer and the trainees who, in many cases, will display cognitive dissonance in disputing the trainer's statements on a particular matter.

Conservatism

Most people are conservative in their views and become increasingly resistant to change. The degree of conservatism is a specific feature of the individual and his attitude to, for instance, safe working procedures, and may be an important barrier to attitude change. Endeavouring to modify behaviour may impinge on a person's conservatism resulting in stress.

Past experience

Most people carry out tasks at work on the basis of past experience, using working practices they have acquired over many years. The majority of people learn by their mistakes and modify their behaviour in order to prevent repetition of mistakes.

Level of intelligence and education

These factors are important in framing attitudes. Well-educated and intelligent people may consider, for example, that instructions from management are only directed at those with poor intelligence or a lack of education and that they do not apply to themselves. Hence there is a need for employers to produce instructions and directions which match the level of intelligence of the workforce. Treating employees as being less intelligent can be stressful for those employees and creates dissension.

Motivation

Motivation, the element of human behaviour that drives people forward, can have a direct effect on attitudes. Bonus and incentive schemes, or various forms of 'planned motivation', are commonly directed at improving the motivation of employees towards improved levels of performance, thereby reaping the rewards offered if targets are achieved. Failure to meet targets, for a range of reasons, can be stressful for the employees concerned and produce resentment amongst those employees.

Credibility
Attitudes are directly associated with the personal beliefs held by people. On this basis, people are more likely to consider changes in attitude suggested if the person making these suggestions has some form of credibility within the organization. Credibility can be associated with a person's rank, position in the organization or level of prestige held. People are more likely to consider changes in attitude to working procedures if the messages are seen to be coming from senior management sooner than from someone further down the organization, such as a supervisor.

Irrespective of whom the message comes from, however, where the message conflicts with certain personal beliefs, stress can be created.

Attractiveness
Any change in attitude requested by senior management must be attractive to the people concerned. Attractiveness, fundamentally, incorporates three elements, that is, similarity, friendship and liking. The more similar two people see themselves in terms of, for instance, a particular work situation, the more likely they are to believe each other. Friendship is also an important factor in attitude change in that a person is more likely to take notice of a friend than someone who is perceived as hostile. Similarly, where a person actually likes and gets on well with, for example, their manager, they are more likely to change their attitude to a situation or work procedure.

Necessary changes in attitude which are perceived as unattractive create a range of stress responses, including anger, frustration and rejection of the changes required.

Selective interpretation
There is an old saying, 'People only hear what they want to hear!' Whether or not a message gets through to bring about change in attitude in the desired direction depends, in large measure, on how the recipient interprets the message. In many cases, the recipient may interpret the facts presented by selecting those that fit in with his existing attitude to that matter. This is especially the case where there is a large discrepancy between the recipient's attitude and the message. Selective interpretation of facts is one of the potent factors in freezing attitudes so that they resist any form of change.

Immunization
This aspect can be compared with the medical practice of immunizing a person against a particular disease by inoculating him with a controlled dose of the organism that causes the disease. In the case of attitudes, the analogy means that a mild exposure to an opposing attitude can immunize a person against it so that he will never accept further facts or arguments for it, no matter how strong they are.

Attitudes currently held
Cognitive dissonance
The problem of cognitive dissonance is a well-established barrier to attitude change and must be considered in the design of, for instance, training programmes and other activities directed at bringing about changes in attitude.

Self-image

Attitudes currently held are an important element of a person's self-image. Similarly, self-image may be identified by a specific set of attitudes. In many cases, a change in a person's attitude to a particular matter may result in loss of face, credibility or self-image, resulting in stress.

Financial gain

The prospect of financial gain can have a significant, but only short-term, effect on attitudes held. People are known to modify behaviour if some form of reward is offered but, on subsequent removal of the reward, commonly revert to their former attitudes. In some cases, the extent of behaviour modification required to achieve the reward is too great, and people who cannot achieve the level of behaviour modification necessary suffer stress in various ways.

The opinions of others

People generally value the opinions of their immediate work contacts, such as managers, trade union representative or other employees. Compliance with these opinions may require some form of behaviour modification which can be stressful.

Skills available

Everyone holds a particular or personal repertoire of skills. The skilled operator frequently holds the attitude that his skills, acquired over many years and, perhaps, through serving an apprenticeship, will automatically protect him against adverse situations, such as, for example, accidents. Similarly, his skills are a source of personal pride that set him above others who do not possess these skills.

Changes in work practices, however, such as 'de-skilling' and the automation of certain processes formerly undertaken by specifically skilled operators, has resulted in loss of status for the skilled operator. This can be particularly stressful for these groups of persons who may feel that they are undervalued in the organization.

The situation

Group situations

People working together on a regular basis tend to develop a group attitude that may be inappropriate. Any training directed at changing attitudes must, therefore, take place on a group basis. Any attempt to break up the group attitudes held must recognize the fact that this may result in stress for members of the group.

The influence of change agents

Employees encounter a number of different 'change agents' at work. These include managers, such as human resources managers, health and safety practitioners and officers of insurance companies. HSE inspectors, environmental health officers and fire prevention officers may bring about change through a range of enforcement activities.

All these people impact on the organization at different levels, in many cases creating stress for the individuals with whom they come into contact.

Prestige

Increasingly, organizations are seeing areas such as quality, health and safety and environmental protection, and the procedures they have adopted, as a means of gaining prestige in the market place and promoting the image of the organization to the outside world. This has resulted in a range of policies and procedures which must be adopted by people further down in the organization who, in turn, may consider that they simply do not have the time to deal with these 'fringe' elements of work, are not interested in these matters or see them as something imposed from above and over which they have no control.

This can result in adverse responses from all levels of the organization and can be a significant cause of stress, particularly to line managers.

The climate for change

Generally, people do not like change of any sort. However, most organizations go through periods of change which can be stressful for those involved at all levels.

Management example

This is the strongest of all motivators for changing attitudes. It may include, when visiting working areas, the simple wearing of certain items of protective clothing by senior managers where the organization is requiring employees to wear similar protective clothing during their work. Poor senior management example, on the other hand, results in their loss of credibility and a certain amount of stress for line managers who may suffer stress as a result of adverse comments made by employees.

Appeals to fear

Attempting to change attitudes through fear is not a good idea. Classic examples of this technique are seen in the 'Don't drink and drive' poster of years ago which showed the mutilated face of a driver who had been projected through the windscreen of a car. The impact was negligible as many people simply 'switched off' when faced with this sort of horror poster. In some cases, however, the response was quite stressful which was one of the reasons why the poster was eventually withdrawn.

6.5.2 Motivation

The term 'motive' implies a need and the direction of behaviour towards a goal, aim or objective. A 'motivator', on the other hand, is something that provides the drive to produce certain behaviour or to mould behaviour.

'Motivation' is a term used to describe the goals or objectives that a person endeavours to meet and the drive or motivating force that keeps him on track in pursuing these goals or objectives. 'Motivation' can be used to refer to states within an individual, to behaviour and to the goals towards which behaviour is directed. Motivation has three specific features:

1. A motivating state within the individual.
2. Behaviour aroused and directed by this state and
3. A goal or objective towards which the behaviour is directed.

When the objective is achieved, the state that caused the behaviour subsides, thus ending a cycle until the state is aroused again in some way.

Theories of motivation

A number of theories of motivation have been developed in the last century in particular those directed at identifying why people work.

F.W. Taylor (1911): Theory of the Management and Organization of Work

Taylor examined the differences between potential managers, who were involved in the organizing, planning and supervision of work activities, and the rest of the workforce, who were not particularly interested in these aspects. They preferred to have simple tasks organized for them that they could be trained to carry out without having to make decisions. Once the work had been planned, money was the principal incentive for increasing production.

Taylor said 'Man is a creature who does everything to maximize self-interest'. In other words, people are primarily motivated by economic gain. 'What's in it for me?' is seen as an important motivator for some people.

This viewpoint, that people only went to work to earn wages, was held by many of the entrepreneurs of the nineteenth and early twentieth centuries. The philosophy that the more money a person was paid, the harder he would work, led to a whole range of variations on this theme, such as incentive schemes, productivity bonuses, piece work, payment by results and even the concept of 'danger money' in certain high risk industries. Taylor's concept (1911) of *Scientific Management* brought about basic strategies and systems, such as the division of labour, mass production processes, work study and great emphasis on the optimum conditions for work, training and the selection of employees.

Taylor's theory has now largely been discredited. Whilst most people would agree that the financial rewards are significant in order for people to survive and maintain a reasonable standard of living, other factors, such as the working environment, social relationships at work, job satisfaction and self-fulfilment are just as important.

Elton Mayo (1930s): Social Man

Various studies by Elton Mayo at the Hawthorn Works of the Western Electric Company, Chicago in the 1930s endeavoured to identify the best working conditions for people with a view to improving productivity. Former attempts to improve worker performance and thereby increase output were based on notions that workers should be regarded as machines, whose output could be improved by attention to the physical conditions and the elimination of wasteful movements and the resulting worker fatigue.

Mayo's methods involved studying the output of a small group of female workers who were consulted about each of a large number of changes to working practices introduced.

Nearly all these changes led to improvements in output. However, when these continuing changes eventually returned to the original physical conditions, output increased still further. Such revelations led Mayo to conclude that, as the changes observed could clearly not be attributed to the changes in the physical conditions as such, they must be due to social factors and to a change in the attitudes of the workers. Fundamentally, because someone was showing an interest in the problems of these workers, they experienced a feeling of importance and responded accordingly by giving their best. It was found, for instance, that most grievances raised were only symptoms of a general discontent, and that a chance to discuss these grievances in the presence of a sympathetic listener frequently led to solving them.

The need for their co-operation, and the fact that consultation had taken place over changes, resulted in the formation of a group culture. Members valued being in the group, felt responsible to the group and the company and, in particular, were concerned with meeting the standards of performance established by the group. The outcome of these studies was the realization that people were not directly motivated by the financial rewards, but tended more towards a social philosophy of 'a fair day's work for a fair day's pay'. They were particularly concerned with the social interaction that took place during working periods and took pride in belonging to an identifiable work group. These factors alone provided a high level of job satisfaction, the more significant the work group, the greater the satisfaction. They also responded to pressure from their peers and the interest shown in them individually by various levels of management. On this basis, work was perceived as a social activity, not just a means of obtaining pay.

A number of important points emerged from Mayo's studies into worker attitudes between 1927 and 1932, in particular:

- The opportunity to air grievances has a beneficial effect on morale.
- Complaints were not often objective, but symptomatic of a more deep-rooted disturbance.
- Workers are influenced by experiences outside their workplace, as well as those inside the workplace.
- Dissatisfaction was often based on what the worker saw as an underestimation of their social status in the firm.
- Voluntary social groups formed at the workplace had a significant effect on the behaviour of individual members giving rise to group norms for production and group sanctions for people who strayed outside these norms or established informal standards.

Mayo's studies further resulted in a changed emphasis on the role of the supervisor, from that of the autocrat to group leader, and a greater attention to the need for establishing and maintaining the morale of work groups.

Abraham Maslow (1943): Self-Actualizing Man

Maslow studied the factors significant in the motivation of successful people and which gave them satisfaction in their work. Achievement, self-esteem and personal growth featured strongly in the outcome of these studies.

Maslow defined motivation as 'a striving to satisfy a need'. Thus, when a person needs something, he will strive to satisfy this need. As a need is satisfied, it uncovers another level of needs, so people go on wanting. Maslow actually categorized these needs and ranked them in order of importance, producing a 'hierarchy of needs' thus:

- **Basic or physiological needs** such as those for food, drink, air, water, sleep, sex and warmth
- **Safety and security needs**, namely the needs for freedom from physical or mental attack, deprivation, insecurity and want; the need for physical and psychological safety and security, and for shelter
- **Social needs**, those needs associated with status and acceptance as a member of the group, that is, a sense of belonging, participation in social activities, self-esteem, respect from one's peers, the need to love and be loved
- **Self-esteem (ego) needs**, the need for achievement, praise, honour, glory and acclaim, the need to be noticed, the need for self-respect, status and the respect of others, for competence, knowledge, independence, responsibility, self-confidence and prestige
- **'Self-actualization'**, the ultimate stage of achieving one's potential; 'doing your own thing'; 'what a man can be, he must be!'; self-actualization is synonymous with growth, personal development, accomplishment, self-fulfilment, self-expression and creativity.

Maslow stated that, at any time, people are progressing from their basic needs to a higher level of need or acting according to a current level of need. When, and only when, a lower need has been satisfied, the next highest becomes dominant and the person's attention is directed to satisfying the higher need. At the higher level, the need for 'self-actualization' is destined to remain unsatisfied as new meanings and challenges arrive. All the above categories of need remain permanently active. What varies is the strength or degree of influence of each of these needs at any one time.

Maslow's original hierarchy of needs is shown in Figure 6.1.

The principal outcome of Maslow's studies is the fact that all needs must be studied in an attempt to motivate people at work. Attempts to provide motivation through the use of incentives relevant to the needs of a lower or higher level, and not that need which is currently dominant, are likely to fail. Furthermore, individuals and groups differ with regard to needs that are dominant. On this basis, the same incentives cannot be expected to motivate everyone, so management should endeavour to establish the dominant needs and act accordingly.

Modifications to Maslow's model

Maslow's original hierarchy of needs model was subsequently modified in the 1970s and 1990s. These modified models are shown in Figures 6.2 and 6.3.

Maslow further identified a number of 'self-actualizing' characteristics, namely:

- A keen sense of reality, aware of real situations, objective judgement rather than subjective judgement
- See problems in terms of challenges and situations requiring solutions, rather than as personal complaints or excuses

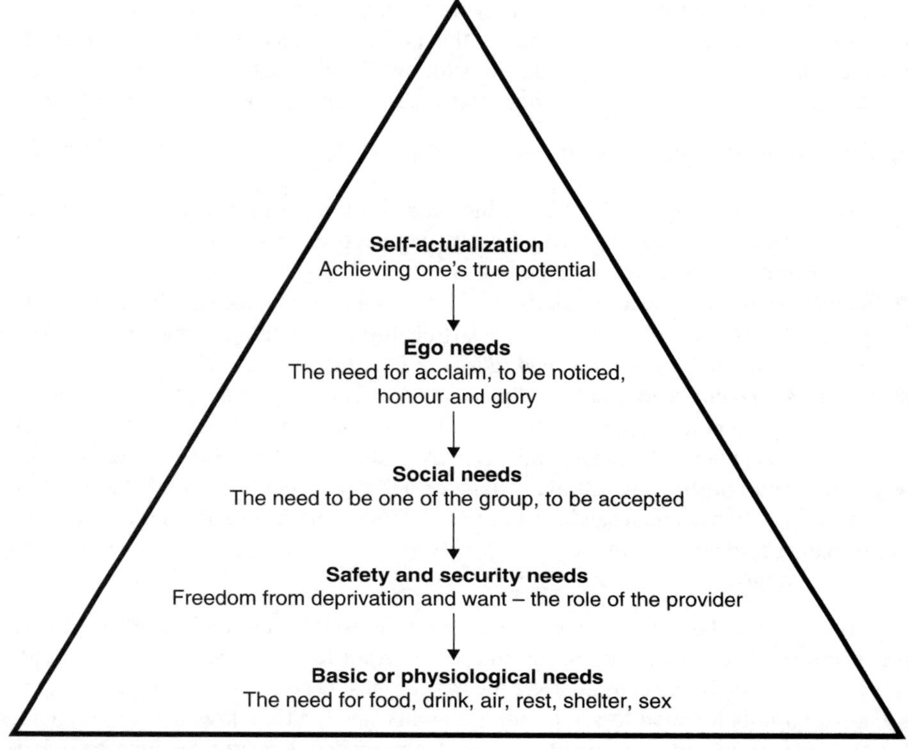

Figure 6.1 Maslow's hierarchy of needs.

- Need for privacy and comfortable being alone
- Reliant on own experiences and judgement, independent, not reliant on culture and environment to form opinions and views
- Not susceptible to social pressures, non-conformist
- Democratic, fair and non-discriminating, embracing and enjoying all cultures, races and individual styles
- Socially compassionate, possessing humanity
- Accepting others as they are and not trying to change people
- Comfortable with oneself; despite any unconventional tendencies
- A few close intimate friends, rather than many surface relationships
- Sense of humour directed at oneself or the human condition, rather than at the expense of others
- Spontaneous and natural, true to oneself, rather than how others want
- Excited and interested in everything, even ordinary things
- Creative, inventive and original
- Seek peak experiences that leave a lasting impression.

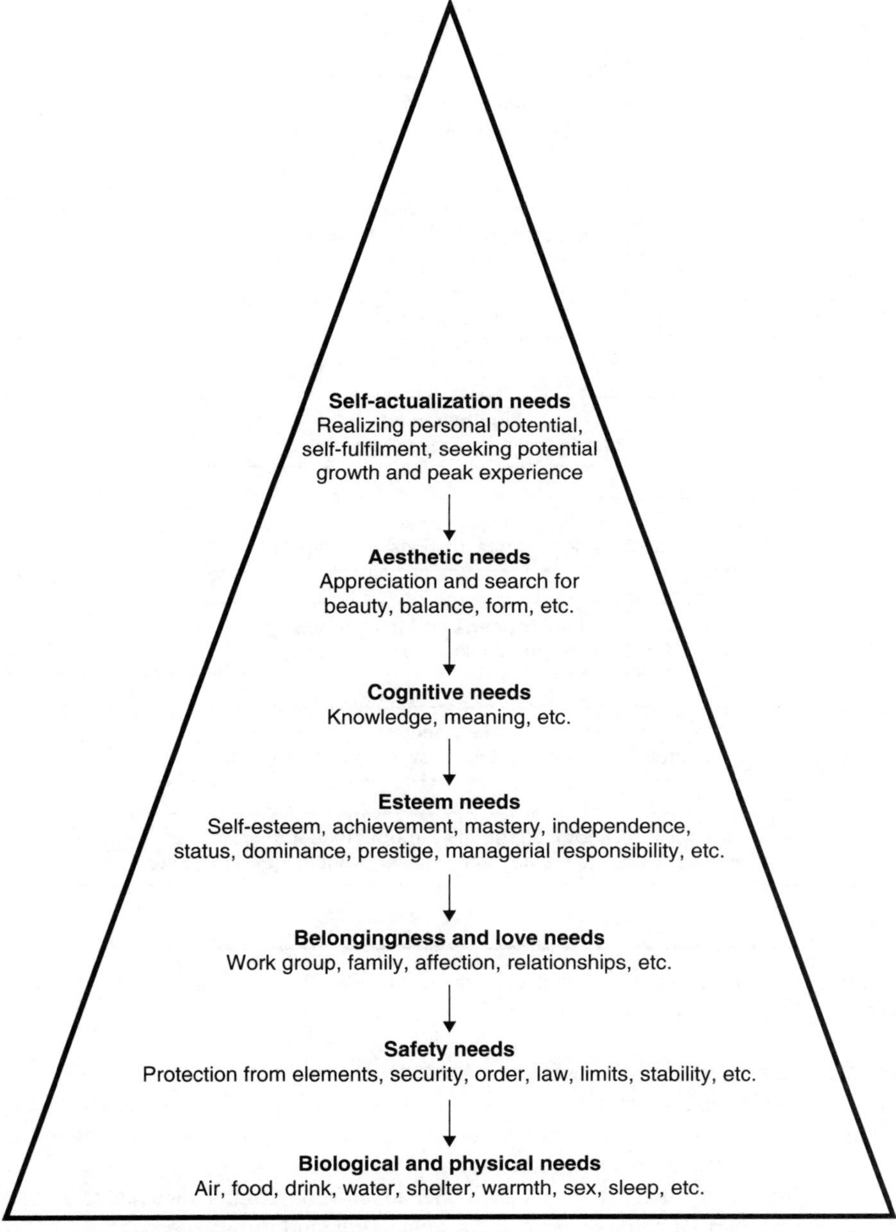

Figure 6.2 Adapted hierarchy of needs model (1970s).

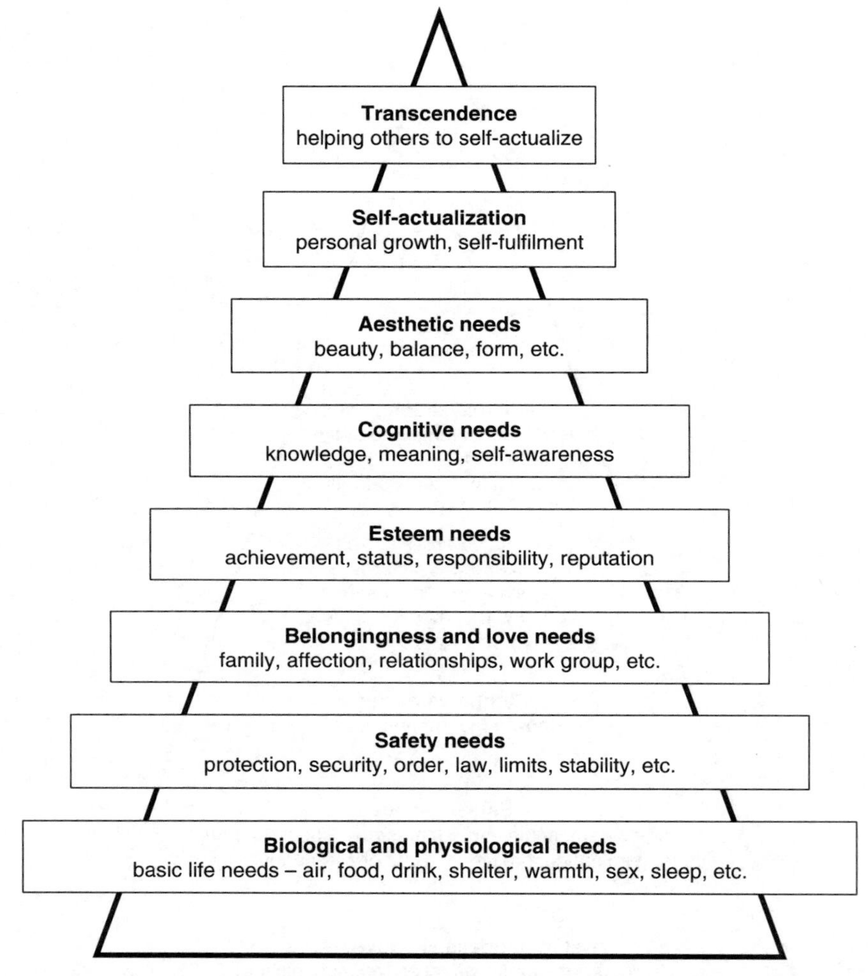

Figure 6.3 Adapted hierarchy of needs model (1990s). *Source*: Alan Chapman, 2002, www.businessballs.com

Fred Herzberg (1957): Two Factor Theory

Herzberg undertook an extensive study throughout the USA, Canada and the United Kingdom seeking to identify among workers in many organizations the factors that produced job satisfaction and dissatisfaction.

He examined the homeostatic needs (hygiene factors or maintenance factors) which are concerned with avoiding pain and dissatisfaction, and growth needs (motivators) concerned with actually achieving satisfaction and fulfilment.

He asked many people in different jobs at different organizational levels two questions:

1. What factors lead you to experience extreme dissatisfaction with your job?
2. What factors lead you to experience extreme satisfaction with your job?

Table 6.1 Hygiene factors and motivators (Herzberg)

Hygiene factors	Motivators
Money	Challenge
Working conditions	Responsibility
Safety arrangements	Advancement
Quality of supervision	Interest and stimulation created by the job
Administrative procedures	Achievement
Interpersonal relationships	Recognition
Status	Possibility of growth
Security	

He established that there was no one factor that determined the presence or otherwise of job satisfaction. However, he did point out that before satisfaction with work can be improved, the factors that cause dissatisfaction must be dealt with.

In a typical work situation maintenance factors involve the total environment affecting the employee – physical conditions, pay, safety, security, social factors and interpersonal relationships. On the other hand, motivational needs that lead to positive happiness are the needs for growth, achievement, responsibility, accountability and recognition. These needs can only be met by undertaking the actual work itself. Herzberg's view was that the job itself can provide a potentially more powerful motivator than any externally introduced motivators. Employees can be actively satisfied only when the work done is perceived by the worker as being meaningful and challenging, thereby fulfilling their motivational needs.

These *hygiene factors* and *motivators* are listed above in Table 6.1.

People expect the hygiene factors to be present and satisfactory. If absent, or poorly managed, they will bring about dissatisfaction. On the other hand, while motivators give rise to job satisfaction, they will not result in dissatisfaction if absent.

The message that comes out of Herzberg's studies is that if management is to motivate people to take greater responsibility and stimulate job interest, they must get the hygiene factors correct, including health and safety arrangements, first. This includes strategies for reducing stress at work.

Improving motivation: Job enrichment

Herzberg suggested the concept of *job enrichment* as a solution to the problem of meeting motivational needs. Care must be taken, however, to ensure that job enrichment techniques are not introduced in such large amounts or at such speed as to arouse excessive alarm, fear amongst workers and stress. When introduced on a planned basis, and with consultation, job enrichment makes it possible for the growth and achievement needs of employees to be met as a result of their efforts at work.

The significance of the job enrichment concept lies in the clarity with which it focuses attention on the motivational distinction between 'task and environment' and 'intrinsic and extrinsic factors'.

Task impoverishment, namely removing the individual interest, challenge and responsibility from a job, results in deteriorating motivation and, for some people, can be a significant cause of stress. Conversely, no amount of environmental improvement, such as improved supervision, working conditions, pay and human relationships, can compensate for this task impoverishment. Clearly, therefore, there is a need to examine the tasks that people do with a view to identifying the factors that provide interest, challenge and responsibility for employees. These higher levels of motivation can be achieved through job enrichment, job enlargement and job rotation accompanied by various worker participation schemes as outlined below:

1. **Job enrichment**: an increase in satisfaction and the responsibility attached to a job is achieved either through reducing the degree of supervision or by allocating each worker a unit of work in which they have freedom to select their method and sequence of operations.
2. **Job enlargement**: in this case, the worker is required to progressively increase the actual number of operations he undertakes.
3. **Job rotation**: while the concept of job rotation has, in some instances been unpopular with workers, the objective is to give more variety on simple, repetitive and usually automated tasks.
4. **Worker participation**: worker participation in varying degrees can be achieved in the following areas:
   - **personnel-related decisions**: transfers to other jobs, disciplinary matters, various forms of training and instruction
   - **social decisions**: welfare arrangements, health and safety procedures and systems of work, regulation of working hours, rest periods
   - **economic decisions**: methods of production, production planning and control, production times, rationalization, changes in plant organization, expansion or contraction.

The fundamental objective is to reduce or eliminate the authoritarian approach to management and to replace it with a more participative style of management. The participative leader is one who plans work and consults his subordinates as to the best course of action. He is skilled in reconciling conflicts so as to achieve group cohesion and effectiveness. Such a person is also interested in individual employees and their particular problems, whereas the authoritarian uses rewards and punishment of the traditional sort, exercises close supervision and is more interested in the activities of those above him rather than below him.

Motivation and stress

Herzberg said 'Before you can improve satisfaction with work, you must remove dissatisfaction with work'. For many people, stress is one of the greatest 'dissatisfiers' with work, in many cases resulting in absenteeism, lack of interest, low self-esteem and poor levels of personal achievement.

Employers should recognize the fact that people go to work not purely for the financial rewards being offered. There are many other work factors which give them satisfaction, happiness, stimulation and interest which they would not get if they stayed at home. Employers, therefore, need to concentrate on these factors causing satisfaction if they are to see improvements in performance. Failure to consider these factors will have the reverse effect characterized by evidence of stress amongst employees.

Three most important aspects in motivating employees and preventing stress are the style and form of communication, the system for consultation and evidence of firm leadership from the top downwards. Consultation with employees in planning the work organization is one of the greatest motivators. The communication process may also need examination with a view to improving vertical and horizontal lines of communication. Management style must incorporate strong leadership and commitment.

The right motivation

Employers should answer the questions below.

1. Do managers at all levels discuss the work with employees?
2. Do managers have an 'open door' policy, making themselves available to discuss the personal problems of employees?
3. Are employees given the chance to discuss personal grievances with managers?
4. Is there evidence of group cultures amongst different work groups?
5. Are people proud to be members of these work groups?
6. Is there a high level of job satisfaction amongst employees?
7. Does the organization consider job enrichment important?
8. Is every job seen as important to the success of the organization?
9. Do supervisors act as group leaders, sooner than autocrats?
10. Are individual achievements in respect of performance targets recognized and rewarded?
11. Are efforts made to boost people's self-esteem?
12. Are efforts made to encourage the personal growth and success of employees?
13. Do managers study the individual needs of employees and endeavour to satisfy these needs?
14. Are employees encouraged to view work problems as challenges and situations requiring a solution?
15. Do you attempt to make people's jobs meaningful, important and challenging?
16. Is there a recognized system for worker participation?
17. Do people get the chance for social interaction at work?
18. Are employees given the chance to improve their knowledge and standing through the provision of information, instruction and training?
19. Are the causes of individual absenteeism investigated?
20. Are the causes of human error investigated?

Ideally, there would be a 'Yes' response to all the above questions.

The quality of working life

In 1974 the International Labour Office Conference passed a resolution directed at promoting and ensuring:

> *protection against physical conditions and dangers at the workplace and its immediate environment; adaptation of installations and work processes to the*

physical and mental aptitudes of the worker through the application of ergonomic principles; prevention of mental stress due to the pace and monotony of work, and the promotion of the quality of working life through amelioration of the conditions of work, including job design and job content and related questions of work organization; the full participation of employers and workers and their organizations in the elaboration, planning and implementation of policies for the improvement of the working environment.

The quality of working life philosophy is one that takes the various theories of motivation and worker satisfaction and places them in a wider framework. It considers the numerous factors that affect the basic satisfaction and, hence, the motivation of employees, and comprise the quality of their life at work. It emphasizes the importance of consultation and worker participation as integral features for improving the quality of working life.

It is generally recognized that a well-satisfied and motivated worker is less likely to contribute to accidents to himself or fellow workers or to suffer stress by virtue of the nature of the work. Moreover, the potential for human error, which could result in accidents or the manufacture of a defective product, is greatly reduced. The provision and maintenance of a working environment free from various forms of environmental stress, such as noise, inadequate illumination and ventilation, or poor welfare facilities, is an important factor in the quality of working life.

6.5.3 Personality

Personality is frequently related to the way people behave. They may be perceived by other people as, for example, rigid, overbearing, honest, flexible or boring. Personality involves the total pattern of behaviour that is unique and manifest in an individual's values, beliefs, interests, attitudes, expressions and actions.

Personality involves individual traits, ways of adjustment, defence mechanisms and ways of behaving that characterize an individual and his relationships with other people, events and situations.

Allport (1961) defined personality as 'the dynamic organization within the individual of those psychophysical systems that determine his characteristic behaviour and thought'. The term 'dynamic' implies that personality is composed of interacting parts and that this interaction produces flexibility of response, that is, an individual who is subject to change. This degree of subjection to change is important, particularly in the selection of people who may be exposed to continuing forms of danger. The term 'psychophysical' means that personality contains both mental and physical elements. 'Determine' in the definition above refers to the notion that personality is considered to be a cause of behaviour.

Personality has been further defined as 'the distinctive characteristics of individuals, the stable and changing relationships between these characteristics, the origins of the characteristics, the ways in which they help or hinder the interaction of a person with other people, and the characteristic ways in which a person thinks about himself'.

The structure of personality

Personality fundamentally comprises a series of traits. A trait is a more or less stable and consistent disposition of an individual to respond to people and situations in a characteristic way – friendly, gregarious, hostile, reserved, humble and so on. Traits provide what is probably the most useful means of characterizing a person.

Personality is directly connected with:

- abilities, attitudes and interests
- motives
- modes of adjustment and
- defence mechanisms.

Adjustment refers to the process of accommodating oneself to circumstances and, more particularly, to the process of satisfying needs or motives under various circumstances. *Defence mechanisms* are important adjustment techniques. A defence mechanism is a device, a mode of behaviour, that a person uses unconsciously and automatically to protect himself against fear, anxiety or feelings of worthlessness that are the emotional consequences of frustration of a motive. Everyone has his own distinct pattern of defence mechanism and, as such, they are important traits of personality.

Personality factors

The best known approach to personality is that produced by R.B. Cattell who considered responses to questionnaires from people on their various beliefs and preferences. From this research Cattell produced a list of 16 personality factors (see Table 6.2 below). These factors are shown as 16 dimensions for which a person's level in each case can be measured and recorded to produce a personality profile, which is unique to the individual completing the questionnaire.

Whilst any clear link between individual personality and an individual's potential for stress is hard to identify, certain personality traits could well be a contributory factor in the level of stress suffered by individuals.

Personality and stress

Could it be deduced that certain persons, with particular personality traits, may be more vulnerable to stress than others? People scoring high in certain elements of the 16 factor test, such as 'reserved', 'affected by feelings', 'humble', 'sober', 'conscientious', 'shy', 'tender-minded', 'apprehensive', 'group-dependent' and 'tense', may well be more vulnerable to stressful events at work, such as petty conflicts over who does what, changes in work practices, together with anti-social and insensitive behaviour by colleagues. This results in worry, anxiety and, for some people, long-standing depression.

Employers should take account of specific personality traits in the allocation of tasks with a view to reducing stress on the persons demonstrating those traits.

A model of personality and stress

Over the last 50 years a number of models have been produced dealing with the relationship of personality to stress, together with numerous self-tests that gave people valuable insights into their own inner workings. One model that is particularly

Table 6.2 Cattell's 16 personality factors

Description Low score	← 1 to 10 →	Description High score
1. Warmth RESERVED, detached, critical, aloof		OUTGOING, warm-hearted, easy-going, participating
2. Reasoning LESS INTELLIGENT, concrete-thinking		MORE INTELLIGENT, abstract-thinking, bright
3. Emotional stability AFFECTED BY FEELINGS, emotionally less stable, easily upset		EMOTIONALLY STABLE, faces reality, calm, mature
4. Dominance HUMBLE, mild, accommodating, conforming		ASSERTIVE, aggressive, stubborn, competitive
5. Liveliness SOBER, prudent, serious, taciturn		HAPPY-GO-LUCKY, impulsively lively, gay, enthusiastic
6. Rule consciousness EXPEDIENT, disregards rules, feels few obligations		CONSCIENTIOUS, persevering, staid, moralistic
7. Social boldness SHY, restrained, timid, threat-sensitive		VENTURESOME, socially bold, uninhibited, spontaneous
8. Sensitivity TOUGH-MINDED, self-reliant, no-nonsense		TENDER-MINDED, clinging, over-protected, sensitive
9. Vigilance TRUSTING, adaptable, free of jealousy, easy to get along with		SUSPICIOUS, self-opinionated, hard to fool
10. Abstractedness PRACTICAL, careful, conventional, regulated by external realities, proper		IMAGINATIVE, wrapped up in inner urgencies, careless of practical matters, Bohemian
11. Privateness FORTHRIGHT, natural, artless, unpretentious		SHREWD, calculating, worldly, penetrating
12. Apprehension SELF-ASSURED, confident, serene		APPREHENSIVE, self-reproaching, worrying, troubled
13. Openness to change CONSERVATIVE, respecting established ideas, tolerant of traditional difficulties		EXPERIMENTING, liberal, analytical, free-thinking
14. Self-reliance GROUP-DEPENDENT, a 'joiner' and sound follower		SELF-SUFFICIENT, prefers own decisions, resourceful
15. Perfectionism UNDISCIPLINED, self-conflict, follows own urges, careless of protocol		CONTROLLED, socially precise, following self-image
16. Tension RELAXED, tranquil, unfrustrated		TENSE, frustrated, driven, overwrought

useful, firstly, in assisting people manage stress and, secondly, in understanding the defence mechanism people present to the world when under stress, is the Enneagram. This term was introduced many years ago, by G.I. Gurdjieff, who was a pioneer in adapting many Eastern spiritual teachings for use by westerners. The approach is based on ancient Sufi teachings that specify nine different personality types and their relationships.

Whilst this is a very broad subject of study, the various personality types are outlined below, together with how people identifying with a particular type respond to stress, and the measures individuals can take to handle stress more effectively.

The perfectionist

The perfectionist says that there is a perfect solution to every situation and if he does not already know this perfect solution, he only has to look for it. Staunchly independent, perfectionists know that hard work and common sense will inevitably lead to success. They are ruled by a harsh internal critic who constantly judges and assesses their actions, telling them to work harder and that they can be successful if they really try.

When things go wrong, a perfectionist will automatically blame himself and, when the criticism becomes too harsh, becomes resentful of those who appear to break the rules without compunction. Their anger, normally turned in on themselves, can explode towards others when perfectionists are convinced they are right.

Perfectionists can handle their stress more effectively by:

- learning to express their anger in acceptable ways;
- questioning the difference between what should be done and what needs to be done;
- modifying their own internal standards to become more open to solutions other than their own one 'correct' way.

The giver

Givers are principally concerned with relationships and firmly believe that if we would only adapt ourselves to the needs of others, the world would be a far better place. Givers are always there when people need a friend, someone to discuss a problem with or a shoulder to cry on. However, givers are easily upset if they are rejected or do not receive the appreciation they feel should be forthcoming. In many respects, they are the Jewish mother stereotype. They can, under stress, become highly manipulative and persecute those who have thwarted them.

Givers can deal with their stress better by:

- recognizing their own tendencies to become helpless and to avoid confusion;
- learning to say 'No' and sticking with that decision;
- detecting the urge to manipulate;
- avoiding histrionic outbursts.

The performer

Performers believe in keeping their days packed with activities and often project the type of image society would appear to value, i.e. successful career, beautiful home and intelligent children. They enjoy thus power over others offered by certain leadership positions and can adapt to almost any audience or situation. Problems arise, however, when the performer confuses a high state of doing with the inner self. In fact, performers commonly get cross with people who fall so easily for what performers know is an act, rather than for whom they really are. Performers feel most alive when in the midst of intense activity. It is the slowing down that causes them the most stress.

In particular, any loss of face or prestige, such as being made redundant, being overlooked for promotion or even being late for an appointment, can be extremely stressful, in many cases leading to depression.

Performers should handle their stress by:

- learning to slow down, or even stop, which leaves room for real feelings and emotions to emerge;
- avoiding the temptation to blame others for failure and to view other people as lazy or incompetent;
- recognizing the differences between doing and feeling, and becoming adept at naming their true feelings;
- accepting that it is perfectly all right if the world does not view them as perfect and that the trappings of success are not the same as true success.

The romantic

Romantics are artistic types who often live at the extremes of emotional reactions and always appear to be suffering. They tend to avoid focusing on the present and idealize whatever is missing in their lives, such as the holiday they never took, the friends they never see or the house they previously lived in. Romantics view themselves as rather special or unique individuals who are capable of more intense emotions than other people. Sadly, they commonly experience deep and dark depression, often triggered by feelings of abandonment. A frequent response to stress is intense self-criticism combined with biting sarcasm intended to punish those who are responsible for their suffering.

Romantics would find their lives less stressful if they:

- recognized the self-absorption that can take place during mood swings and try to focus on what is important to others;
- finished projects even when they involved some mundane activity;
- exercised to improve their moods;
- learned to avoid pulling others into their histrionic moments by trusting that their own moods will improve.

The observer

Observers are the most private of all types and, generally, prefer to avoid becoming involved in anything which may disturb their way of life. They are intensely independent and are quite comfortable living alone on a modest income and avoiding life's trivialities. They tend to be good listeners as this helps keep the spotlight off themselves. They avoid emotional entanglements. They prefer meetings with agendas announced well in advance so they can prepare.

However, observers encounter stress in the unexpected and the unpredictable. They do not respond well to change at work or at home, particularly that which is unexpected and suddenly announced. When put under stress, they draw back and often describe themselves as observing life around them as if they were in a tunnel. When challenged by others, particularly those whose values and knowledge of a lower level than their own, they may become irate.

Observers need to:

- cultivate a spirit of spontaneity;
- try to avoid withdrawal as a way to control others;
- resist the temptation to replace emotions with analysis;
- recognize that misstatements by others are not always challenging to their integrity.

The loyal sceptic

Loyal sceptics have deeply rooted fears that can lead in either of two very different directions. They may become phobic or fearful of nearly anything in the environment or counterphobic and directly challenge their worst fears by taking up dangerous pursuits, such as sky diving, or becoming police officers.

Loyal sceptics often have difficulty in dealing with authority. This can be expressed by either unquestioning obedience or, alternatively, rebellion. Their ability to see threats everywhere, if not controlled may blossom into full-blown paranoia. Because they see their role as challenging authority, whether at local management level or government level, they are often called devil's advocates for their willingness to challenge the power structure, but sometimes they are acting on their fear of success.

When under stress, loyal sceptics become mistrustful of nearly everyone, constantly looking for signs that support their suspicions. They can become hyperactive and overly aggressive.

Loyal sceptics would better handle their stress by:

- accepting that some uncertainty and insecurity are a natural state and learning to live with them and working to recall positive memories as well as negative;
- recognizing that flight is as much a reaction to threats as is fight;
- checking out fears with others and moving ahead despite the fear;
- reining in their tendency to challenge authority or to accept it unquestioningly.

The epicure

Epicures mask their fear by moving towards others, trying to charm them. Epicures look unafraid and are often liked for their sunny dispositions. They love planning because it lets them keep multiple options open, and their fondness for play earns them the nickname of Peter Pan. However, multiple options can become a way to avoid following through with a course of action that becomes unpleasant. Their sense of fun may come forward at inappropriate times, especially when they are under stress. At these times they may flit from topic to topic, over schedule, launch new projects, anything to avoid dealing with a painful situation. Other typical responses are to paint the other person as the sole source of a problem, rather than looking within or to express anger by making fun of problems in an attempt to trivialize them.

Epicures will better handle their stress by:

- staying with a painful situation long enough to see that a problem exists;
- resisting the temptations to switch to new options rather than dealing with the stressful situation;
- learning to appreciate deeper experiences rather than staying on the surface;
- monitoring the tendency to charm the situation, including fabrication of entertaining stories.

The protector

Only the strong survive and protectors see their role as shielding those who are not as strong as they see themselves as being. They prefer to take charge in a situation because that will let them maintain control. They are often aggressive and feel comfortable, even justified, in expressing their anger. Protectors tend towards extremes; others are weak or strong, situations are fair or unfair, and another's view is right or wrong. When under stress, protectors tend to lock into place and are unable to see anything other than the weaknesses of the opponent. When they feel wronged, they will seek revenge, and they will fight on and on when they believe justice is at stake. Rather than show vulnerability, protectors will blame others and find fault.

Protectors will better handle their stress by:

- resisting the temptation to lecture others, especially when the protector thinks it's for the other person's 'own good';
- learning to compromise, at first as a necessary means to an end and later on as a path to better and fairer solutions;
- recognizing when the need to control is being used as a way to identify friends and foes;
- understanding that not everyone wants to be protected and that their aggressive behaviour can alienate the very people they are trying to protect;
- trying to delay their angry responses by any means possible, e.g. breathing deeply and calmly, counting to 10, learning to listen better.

The mediator

Mediators can see many sides to every issue and are often the peacemakers in a group. They prefer to 'stay on the fence' because to commit would force them to say 'No' to someone. It's much easier to go along with others' wants rather than ferret out their own needs and desires. Despite their seemingly easy-going disposition, mediators possess a stubborn streak stronger than that of any other type. While considering multiple viewpoints, as they take forever to make up their minds, often flatly refusing to be hurried, they are every bit as stubborn about staying with a decision once it's been made. When under stress, they dig in their heels very deeply. Mediators bury their anger very deep within and often express it as passive–aggressive behaviour.

Mediators will better handle their stress by:

- using interim deadlines and other project management techniques to reach the ultimate goal;
- asking for choices, then eliminating the unattractive options to more quickly find their preferences;
- voicing their opinions;
- being aware of when they are 'digging in their heels'.

6.6 Mentally and physically challenged employees

Certain employees may be classed as 'mentally or physically challenged' implying that they suffer some form of disability. They may experience difficulty with undertaking normal workplace activities. As such, they are more vulnerable to stress and may need special consideration by employers in the allocation of tasks. They should be constantly under surveillance by their line manager to ensure they are not suffering from stress through work which is too demanding or challenging. Failure to undertake this task could lead to subsequent civil actions by such employees.

6.7 Workplace indicators of stress

Employers need to recognize the indicators of stress in the workplace. Classic indicators include:

- Decrease in performance amongst specific groups
- Missed deadlines
- Poor levels of punctuality
- An increase in sickness absence
- Evidence of aggression and anger amongst employees
- Increased irritability at management decisions
- A tendency towards withdrawal and evidence of depression
- A reduced interest in work and in fellow employees
- Lack of concentration.

6.8 Stress in groups

6.8.1 The 'flight or fight' response in groups of workers

Selye's 'flight or fight' response can be important in considering behaviour in work groups who are experiencing too much or too little pressure in their work. Groups who demonstrate the 'fight' response can be characterized by:

- A 'blaming' culture
- Punishing members who do not meet work standards
- Being excessively competitive
- Being political in their outlook
- Aggressive behaviour
- The operation of hidden agendas
- Strikes.

The 'flight' response, on the other hand, may show evidence of:

- Withdrawn behaviour by individuals
- Lack of contribution to discussions on the work process
- Submissive attitudes
- Ignoring problems arising from the work
- Subversive activities

- Lack of communication with members of the work group
- High levels of sickness absence.

6.9 The sources of management stress: HSE guidance note HS(G)48

Managers and employees within an organization are open to stress. The HSE (1999) publication *Reducing Error and Influencing Behaviour* [HS(G)48] covers many of the causes of stress and the remedies employers should consider. Stress is also a significant factor in human error. This HSE guidance note, which was originally entitled *Human Factors and Industrial Safety* was revised and updated in 1999. It provides an excellent introduction to the subject of human factors, emphasizing the need to carry out risk assessments which consider human factors and the fact that human factors is a key ingredient of effective health and safety management.

The guide deals with:

- The main types of human failure (slips, lapses and mistakes), and the importance of designing jobs and procedures for the effective control of human errors and violations.
- Methods for improving health and safety through better design of tasks, equipment (ergonomic design), procedures and warnings. This may involve Human Reliability Assessment (HRA) which generally includes the estimation of Human Error Probabilities (HEP) using techniques such as THERP, HEART, SLIM, etc.
- Operational issues, such as shiftwork, fatigue, shift communication, risk perception, influencing behaviour and safety culture.
- Practical measures that organizations can take to ensure that human factors are considered (in risk assessments, accident analyses, design and procurement). This includes checklists for human factors in the workplace.
- Eighteen case studies to illustrate successful human factors interventions.

Inadequate design of plant and equipment is identified as a potential technical failing that can lead to accidents. Designers must always take into account human fallibility and never presume that those who operate or maintain plant or systems have full and continuous appreciation of essential features. The importance of 'fail safe' systems is emphasized.

6.9.1 Key features of human factors

The three key features in considering a human factors-related approach to stress are job or task factors, individual or personal factors and organizational factors.

Job factors

Inadequate and poor design of jobs can be a contributory factor in increasing the stress on employees. Tasks should be designed in a way that takes into account ergonomic principles recognizing strengths and limitations in human performance. Matching the individual to the job must be undertaken carefully and requires consideration of both physical and mental match to the tasks undertaken. The mismatch between job requirements and individual capabilities provides the potential for human error. The need to take into account human capability when allocating tasks is required under the Management of Health and Safety at Work Regulations.

Individual factors

Individual characteristics, such as motivation, attitudes held, skills, habits and personalities, can be strengths or weaknesses depending upon task demands. Certain individual characteristics, such as personality, tend to be fixed, whereas other characteristics, such as skills and attitudes, can be modified or enhanced. Attitudes to safety vary significantly.

Organizational factors

Organizational factors have the greatest influence upon individual and group behaviour. Factors such as commitment, demonstration of intentions and leadership from management are important. The organizational culture, for example, needs to promote employee involvement and commitment at all levels and to emphasize that deviation from established health and safety standards is not acceptable.

6.10 Stress and the potential for human error

Errors are more likely to occur where there are workplace environmental stressors, extreme task demands, social and organizational stressors and equipment-related stressors. Any risk assessment should therefore take these factors into account with a view to identifying the potential for errors arising and their consequences. A programme designed to reduce and control errors should incorporate the following objectives:

- To address the working conditions with a view to reducing stress
- To design plant and equipment in order to either prevent slips or lapses occurring or to increase the chances of identifying and correcting these errors
- To ensure arrangements for training are effective
- To design jobs to avoid the need for tasks which involve complex decisions, diagnoses or calculations
- To provide proper supervision
- To check that job aids such as procedures are clear, concise, available, up-to-date and accepted
- To monitor measures taken to reduce operator error are effective.

6.11 Conclusion

Complaints of stress at work to the enforcement agencies are increasing, particularly in the light of recent civil cases which have laid down the ground rules for proving 'psychiatric injury' arising from stress (see Chapter 8).

Employers need to develop a strategy for dealing with this matter, perhaps commencing with a formal Statement of Policy on Stress at Work. In particular, risk assessments undertaken by employers must take the potential for stress into account and incorporate measures for eliminating or controlling stress. There is a need for organizations to consider the area of human factors and safety with particular reference to the role of the organization, the tasks that people undertake and personal factors, such as attitude, motivation and personality, when allocating tasks.

Questions to ask yourself after reading this chapter

- What strategies should employers use to prevent or reduce stress amongst employees?
- Does your management culture help in preventing stress?
- Are there problems arising from stress in your organization? Have you considered the signs of stress amongst employees?
- Does your organization take a human factors-related approach to managing health and safety?
- Do you ascribe to the theories of motivation outlined in this chapter?
- Do you believe in the concepts of 'job enrichment' and 'job impoverishment'?
- Are some employees more prone to stress on account of certain personality traits?
- Can you recognize certain 'personality types' in your organization?
- What workplace indicators of stress would apply in your organization?
- Are there any work groups who may be more prone to stress than others?
- What are the principal causes of human failure in your organization?

Key points – implications for employers

- According to the HSE, about half a million people experience work-related stress at a level they believe is affecting their health.
- Employers need to develop the right attitude to stress, job factors, management style and systems for dealing with change as important factors in reducing stress.
- Employers need to recognize evidence of stress amongst employees through aspects such as work performance, attitudes and behaviour, relationships at work, error rates and current levels of sickness absence.
- The risk assessment process should consider human factors and human factors must be viewed as an important element of effective health and safety management.
- Stress is a particular cause of human error, a contributory factor in many accidents.

7

Managing stress at work

7.1 Employers' responsibilities and duties in relation to stress

It will be seen from Chapter 9 that every employer has a general duty under section 2(1) of the Health and Safety at Work Act to ensure, so far as is reasonably practicable, the health of all his employees. 'Health' includes not only physical health but mental health. Failure to consider the mental health of employees could result in action by the enforcement agencies, particularly in the light of the increased attention to the subject by the HSE.

The civil implications (see Chapter 8) are even more significant in that it is now well established that stress arising from work can result in psychiatric injury. Recently, this has become the basis for a number of civil claims, with damages payable in six figure amounts. Employers need, therefore, to develop a strategy for countering stress-induced injury amongst employees.

A number of strategies can be considered in dealing with work-related stress, namely:

- recognizing evidence of stress
- prevention of stress
- management of stress
- rehabilitation.

7.1.1 Recognizing evidence of stress

There are numerous manifestations of an organization under stress. These include high levels of sickness absence, poor timekeeping, alcoholism, poor relationships between managers and employees, evidence of bullying and harassment, high staff turnover and poor communication, all of which are covered in other chapters.

7.1.2 Prevention of stress

The old maxim, 'Prevention is better than cure' is significant in this case. One of the starting points in identifying and subsequently preventing stress in employees is through a personal stress audit shown later in this chapter. Personal stress audits (or self-reporting questionnaires) are an important means of identifying common causes of stress, monitoring stress levels and identifying the measures necessary at organizational level to alleviate the problem.

As with any form of monitoring system, feedback and management action following this audit is important. This may take the form of information, instruction and training for management and employees, stress management courses, the development and promotion of a policy on stress at work, perhaps produced as a sub-policy to the Statement of Health and Safety Policy, and encouraging employees to report and discuss stressful elements of their work with their managers. Above all, the organization has to recognize the existence of stress at work and the fact that stress is a common feature of many workplace activities and tasks.

7.1.3 Management of stress

Again, emphasis is placed on education and training. Everyone has their own personal stress response, such as insomnia, loss of appetite or greatly increased appetite, lower back pain, headaches and general fatigue. In this case, employees need advice on identifying their own personal stress response and the measures necessary to coping with it.

Two HSE publications provide excellent guidance on dealing with stress. *Tackling Work-Related Stress: A Guide for Employees* (HSE, 2003) is aimed at employees in all work sectors. It describes work-related stress, personal strategies for dealing with same and measures to take after a stress-related illness. *Tackling Work-Related Stress: A Manager's Guide to Improving and Maintaining Employee Health and Well-being* (HS(G)218) (HSE, 2001) on the other hand, is aimed at managers in organizations employing over 50 employees. It provides practical advice on measures managers can take to assess and control the risks from work-related stress.

7.1.4 Rehabilitation

This third level of action is directed at treating people who have suffered some form of psychiatric injury or mental health problems and can take the form of Employee Assistance Programmes. These programmes incorporate a number of elements, including counselling on stress-related issues in people's lives, setting of personal objectives, including those to promote a healthier lifestyle, measuring stages of improvement in reducing, for instance, anxiety and depression, and prompt referral of affected individuals for specialist treatment.

There is evidence to show that most people developing mental illness will make a complete recovery over a period of time and will, subsequently, return to work.

7.2 Duties of senior management: The human factors-related approach

The last decade has seen increased emphasis on human factors in the workplace. The original HSE publication *Human Factors and Industrial Safety* [HS(G)48] defined 'human factors' as a term used to cover:

- the perceptual, physical and mental capabilities of people and the interaction of individuals with their job and the working environments;
- the influence of equipment and system design on human performance; and
- the organizational characteristics which influence safety-related behaviour.

This guidance document refers to the areas of influence on people at work as being the organization, the job itself and personal behavioural factors. These areas of influence are directly affected by the system for communication within the organization, together with the training systems and procedures in operation, all of which are directed at preventing human error.

This entails examining these areas of influence on people at work with particular reference to the organizational arrangements for dealing with health issues, the design and structure of jobs, and personal factors, such as the attitudes, motivation, personalities and perceptions of individuals. Any examination of this type must take into account the potential for stress.

7.2.1 Organizational characteristics

Organizational characteristics which influence health and safety-related behaviour include:

- the need to produce a positive climate in which health and safety is seen by both management and employees as being fundamental to the organization's day-to-day operations, that is, they must create a positive health and safety culture;
- the need to ensure that policies and systems which are devised for the control of risk from the organization's operations take proper account of human capabilities and fallibilities;
- commitment to the achievement of progressively higher standards which is shown at the top of the organization and cascaded through successive levels of the organization;
- demonstration by senior management of their active involvement thereby galvanizing managers throughout the organization into action; and
- leadership where an environment is created which encourages safe behaviour.

All the above factors feature in the duties of senior management and inadequate attention to them is important in any consideration of the causes of stress at work.

7.2.2 Job design

Successful management of human factors and the control of risk involves the development of systems which take proper account of human capabilities and fallibilities. Tasks should be designed in accordance with ergonomic principles so as to take into account limitations in human performance. Matching the man to the job will ensure that he is not overloaded, and that he makes the most effective contribution to the enterprise.

Physical match includes the design of the whole workplace and working environment. *Mental match* involves the individual's information and decision-making requirements, as well as their perception of tasks. Mismatches between job requirements and workers' capabilities provide potential for human error.

Major considerations in job design include:

- Identification and comprehensive analysis of the critical tasks expected of individuals and appraisal of likely errors
- Evaluation of required operator decision-making and the optimum balance between human and automatic contributions to the safety actions
- Application of ergonomic principles to the design of the man–machine interfaces, including displays of plant process information, control devices and panel layouts
- Design and presentation of procedures and operating instructions
- Organization and control of the working environment, including the extent of the workspace, access for maintenance work, and the effects of noise, lighting and thermal conditions
- Provision of the correct tools and equipment
- Scheduling of work patterns, including shift organization, control of fatigue and stress, and arrangements for emergency operations and situations
- Efficient communication, both immediate and over periods of time.

Inadequate job design is frequently a cause of stress amongst employees, and increases the potential for human error and accidents.

7.2.3 The potential for human error

Limitations in human capacity to perceive, attend to, remember, process and act on information are all relevant in the context of human error. Typical human errors are associated with lapses of attention, mistaken actions, misperceptions, mistaken priorities and, in some cases, wilfulness.

7.3 Human factors and the need to manage stress

Most employers are aware of the fact that a lack of attention to the 'people side' of the organization is one of the principal causes of business failure. Failures in leadership and in demonstrating a caring attitude towards employees results in low morale and lack of commitment by employees. Moreover, stress has a direct association with the perceptual, physical and mental capabilities of people. The concept of matching the man to the job is not only important in ensuring sound levels of performance but in the prevention of stress due to incorrect mental match.

Senior management has the task of driving forward the organization's policy and procedures relating to stress. The above factors should be considered by employers when devising these policies and procedures. It should be recognized that potentially stressful organizations are those:

- which are large and bureaucratic;
- in which there are formally prescribed rules and regulations;
- where there is conflict between people and positions;
- where people are expected to work for long hours;
- where no praise is given;
- where the general conditions are classed as 'unfriendly'; and
- where there is conflict between normal work and outside interests.

Stress prevention is an important feature of the human factors-related approach to management and must feature in policies and systems accordingly.

7.4 Developing a strategy

1. **Recognizing the problem**: Generally, an organization will be more effective if there is conscious recognition of stress potential and efforts are made to eliminate or reduce stress amongst employees.
2. **Morale**: One of the standard criticisms from people at all levels is that the organization does not care about its people. This feeling is reflected in attitudes to management, the job and the organization as a whole. It is important, therefore, for the organization to show at all levels that it really does care. This will result in increased motivation and a genuine desire on the part of staff to perform better. There is clear-cut evidence throughout the world which shows that the most profitable companies are those which take an interest in their staff and promote a caring approach.

7.4.1 The benefits to the organization of reducing stress

The benefits to both employers and employees from reducing stress can be summarized thus:

- Improved health and morale
- Reduced levels of sickness absence
- Increasing levels of performance
- Improved relationships with work colleagues, clients and customers
- Reduced employee turnover
- Reduced employee costs.

7.5 Strategies for managing stress

There is a need here to consider both organizational and individual strategies for managing stress in the workplace.

7.5.1 Organizational strategy

- **Employee health and welfare**: Various strategies are available for ensuring sound health and welfare of employees. These include various forms of health surveillance, health promotion activities, counselling on health-related issues and the provision of good quality welfare amenity provisions, i.e. sanitation, washing, showering facilities, facilities for taking meals, etc.
- **Management style**: Management style is frequently seen as uncaring, hostile, uncommunicative and secretive. A caring philosophy is essential, together with sound communication systems and openness on all issues that affect staff.
- **Change management**: Most organizations go through periods of change from time to time. Management should recognize that impending change, in any form, is one of the most significant causes of stress at work. It is commonly associated with job uncertainty, insecurity, the threat of redundancy, the need to acquire new skills and techniques, perhaps at a late stage in life, relocation and loss of promotion prospects. To eliminate the potentially stressful effects of change, a high level of communication in terms of what is happening should be maintained and any such changes should be well managed on a stage-by-stage basis.
- **Specialist activity**: Specialist activities, such as those involving the selection and training of staff, should take into account the potential for stress in certain work activities. People should be trained to recognize the stressful elements in their work and the strategies available for coping with these stressors. Moreover, job design and work organization should be based on ergonomic principles.

7.5.2 Individual strategy

There may be a need for individuals to:

- develop new skills for coping with the stress in their lives
- receive support through counselling and other measures
- receive social support
- adopt a healthier lifestyle and
- where appropriate, use support from prescribed drugs for a limited period.

The advice of occupational health practitioners is recommended in these circumstances.

7.6 HSE management standards

The HSC has introduced a programme of work to tackle occupational stress through a range of actions, including the development of good standards of management practice. Based on the responses to a former discussion document and the results of a research programme, the HSC concluded that:

- work-related stress is a serious problem
- work-related stress is a health and safety issue and
- it can be tackled in part through the application of health and safety legislation.

However, in the absence of any clear standards of management practice against which an employer's performance in managing a range of stressors, such as the way work is structured, could be measured, the HSC asked the HSE to develop standards of management practice for controlling work-related stressors.

The HSE's strategy on stress has four main themes:

1. To work with partners to develop clear, agreed standards of good management practice for a range of stressors.
2. To better equip HSE inspectors and local authority officers to be able to handle the issue in their routine work, for instance by providing information on good practice and advice on risk assessment and consultation in the light of the above work.
3. To facilitate a comprehensive approach by starting a project, perhaps along the lines of their current *Back in Work* initiative, that will seek to involve others actively in developing a more comprehensive approach to managing stress.
4. To develop additional detailed guidance, drawing on the findings from HSE's research and adopting a particular focus on risk assessment.

The management standards are aimed at those stressors which affect the majority of employees in an organization and cover six main factors which can lead to work-related stress. The standards, and the state to be achieved, are outlined below.

Demands

The organization has achieved the standard if:

- at least 85 per cent of employees indicate that they are able to cope with the demands of their jobs; and
- systems are in place locally to respond to any individual concerns.

State to be achieved:

- The organization provides employees (including managers) with adequate and achievable demands at work.
- Job demands are assessed in terms of quantity, complexity and intensity and are matched to people's skills and abilities.
- Employees have the necessary competencies to be able to carry out the core functions of their job.
- Employees who are given high demands are able to have a say over the way the work is undertaken (see standard on Control, below).

- Employees who are given high demands receive adequate support from their managers and colleagues (see standard on Support, below).
- Repetitive and boring jobs are limited, so far as is reasonably practicable.
- Employees are not exposed to a poor physical working environment (the organization has undertaken a risk assessment to ensure that physical hazards are under appropriate controls).
- Employees are not exposed to physical violence or verbal abuse.
- Employees are provided with mechanisms which enable them to raise concerns about health and safety issues (e.g. dangers – real or perceived, working conditions) and working patterns (e.g. shift work systems, uncertain hours, etc.) and, where necessary, appropriate action is taken.

Control

The organization has achieved the standard if:

- at least 85 per cent of employees indicate that they are able to have a say about the way they do their work; and
- systems are in place locally to respond to any individual concerns.

State to be achieved:

- The organization provides employees with the opportunity to have a say about the way their work is undertaken.
- Where possible, the organization designs work activity so that the pace of work is rarely driven by an external source (e.g. a machine).
- Where possible, employees are encouraged to use their skills and initiative to complete tasks.
- Where possible, employees are encouraged to develop new skills to help them undertake new and challenging pieces of work.
- Employees receive adequate support when asked to undertake new tasks. Employees are supported, even if things go wrong.
- Employees are able to exert a degree of control over when breaks can be taken.
- Employees are able to make suggestions to improve their work environment and these suggestions are given due consideration.

Support

The organization has achieved the standard if:

- at least 85 per cent of employees indicate that they receive adequate information and support from their colleagues and superiors; and
- systems are in place locally to respond to any individual concerns.

State to be achieved:

- The organization provides employees (including managers) with adequate support at work.
- There are systems in place to help employees (including managers) provide adequate support to their staff or colleagues.

- Employees know how to call upon support from their managers and colleagues.
- Employees are encouraged to seek support at an early stage if they feel as though they are unable to cope.
- The organization has systems to help employees with work-related or home-related issues (e.g. EAPs) and employees are aware of these.

Relationships

The organization has achieved the standard if:

- at least 65 per cent of employees indicate that they are not subjected to unacceptable behaviours (e.g. bullying) at work; and
- systems are in place locally to respond to any individual concerns.

State to be achieved:

- The organization has in place agreed procedures to effectively prevent, or quickly resolve, conflict at work.
- These procedures are agreed with employees and their representatives and enable employees to confidentially report any concerns they might have.
- The organization has a policy for dealing with unacceptable behaviour at work. This has been agreed with employees and their representatives.
- The policy for dealing with unacceptable behaviour at work has been widely communicated in the organization.
- Consideration is given to the way teams are organized to ensure that they are cohesive, have a sound structure, clear leadership and objectives.
- Employees are encouraged to talk to their line manager, employee representative or external provider about any behaviours that are causing them concern at work.
- Individuals in teams are encouraged to be open and honest with each other and are aware of the penalties associated with unacceptable behaviour.

Role

The organization has achieved the standard if:

- at least 65 per cent of employees indicate that they understand their role and responsibilities; and
- systems are in place locally to respond to any individual concerns.

State to be achieved:

- The organization ensures that, as far as possible, the demands it places on employees (including managers) do not conflict.
- The organization provides inductions for employees to ensure they understand their role within the organization.
- The organization ensures that employees (including managers) have a clear understanding of their roles and responsibilities in their specific job (this can be achieved through a plan of work).
- The organization ensures that employees understand how their job fits into the overall aims and objectives of the organization/department/unit.

- Systems are in place to enable employees to raise concerns about any uncertainties or conflicts they have in their role.
- Systems are in place to enable employees to raise concerns about any uncertainties or conflicts they have about their responsibilities.

Change

The organization has achieved the standard if:

- at least 65 per cent of employees indicate that the organization engages them frequently when undergoing an organizational change; and
- systems are in place locally to respond to any individual concerns.

State to be achieved:

- The organization ensures that employees (including managers) understand the reason for the proposed changes.
- Employees receive adequate communication during the change process.
- The organization builds adequate employee consultation into the change programme and provides opportunities for employees to comment on the proposals.
- Employees are made aware of the impact of the changes on their jobs.
- Employees are made aware of the timetable for action, and the proposed first steps of the change process.
- Employees receive support during the change process.

7.7 Teamworking

The practice of teamworking has been common in many organizations over the last 50 years. It implies groups of people working together, perhaps as a small production unit or engaged in a particular task. However, a team is more than just a group of people with a common objective working in close proximity. Like a soccer team, members of the group must collaborate with each other, co-ordinating interdependent activities, in order to achieve shared goals or objectives. Teams can be successful if they have a positive impact on factors such as job autonomy, skill variety and feedback. On the other hand, teams that remove job discretion and increase workload are unpopular and cause negative effects with the workforce.

The HSE (2002) report *CRR 393 Effective Teamwork: Reducing the Psychosocial Risks, Case Studies in Practitioner Format* indicates that teamworking has the effect of both increasing and decreasing work-related stress levels. Much depends upon the design of the team and the methods of implementation used by employers. In the first case, there is a danger that teamworking can cause an increase in employee stress levels through an increasing workload and some uncertainty as to what is expected of them in this new system of working. However, where employees have been advised of the purpose and benefits of teamworking, successful teamworking can result in a reduction in work-related stress through enabling greater discretion over their working environment and increasing job challenges.

7.8 Decision-making and stress

Managers and employees frequently make decisions on a range of matters. In many cases, this may entail on-the-spot or snap decisions, some of which may incorporate an element of personal risk-taking. Stress can have a direct effect on this decision-making process, whether it entails operating a process in a particular way, to invest capital in new machinery or to invest in a particular area of the stock market.

Decision-making requires the ability to concentrate on the task, an ability to perceive new information, good short-term memory, planning and time to consider the options. It must be rational, based on sound judgements, correct and up-to-date information and an understanding of the risks involved.

The quality of decision-making can, in many cases, be related to the level of stress imposed on an individual, in other words, the greater the stress, the worse the level of decision-making. The following points need consideration:

- The greater the stress, the greater the likelihood that a decision-maker will choose a risky alternative.
- Groups experiencing substantive conflict more frequently employ creative alternatives to achieve more productive decisions than groups without conflict.
- The greater the group conflict aroused by crisis, the number of communication channels available to handle incoming information decrease.
- During crisis, the ability of the group to handle difficult tasks requiring intensely focused attention is decreased.
- The greater the stress, the greater the tendency to make a premature choice of alternatives for a correct response.
- The greater the stress, the less likely that individuals can tolerate 'ambiguity'.
- Under increasing stress, there is a decrease in productive thoughts and an increase in distracting thoughts.
- The greater the stress, the greater the distortion in perception of threat and poor judgement often occurs.
- The greater the fear, frustration and hostility aroused by a 'crisis', the greater the tendency to aggression and escape behaviours.
- In a stressful situation (whether real or perceived stress), only immediate survival goals are considered which means that longer range considerations must be sacrificed.

Fundamentally, people should not be pressurized to make snap decisions. Decisions made under stressful circumstances are frequently bad decisions with subsequent adverse results.

7.9 The work setting

The 'setting' or framework within which work takes place is important in the prevention of stress. This includes the organization's culture and climate, frequently dictated by management style and behaviour, long-established work customs and practices, specific procedures for dealing with particular issues, decision-making and the environment of work, with respect to, for example, the structure and layout of the workplace, comfort conditions, such as humidity control, the scale of amenity provision and the social interactions that take place between people at all levels. Where the framework of work is 'friendly', as opposed to 'hostile', the potential for stress is much less.

This framework is also concerned with the relative freedom of employees with respect to work content. The potential for work-related stress is less amongst those who are empowered to plan their own work, control their workloads, make decisions regarding the completion of the work and solve their own problems.

7.10 Information, instruction and training

Organizations should have a system for the provision of information, instruction and training for employees on a range of topics. Apart from the fact that the giving of the training is a legal requirement under the HSWA and many regulations made under the Act, there are many benefits to employers in having a well-informed and trained workforce.

Employees should be made aware of the problem of stress through a range of mechanisms, such as the running of stress awareness courses, access to government, HSE and other publications on the subject, the organization's Statement of Policy on Stress at Work and established procedures whereby employees can report stress to an informed manager and receive counselling where appropriate.

7.11 Communicating change

Stress is very much concerned with how people deal with changes in their lives, both at home and at work. One of the principal causes of work-related stress is the failure of the organization to communicate change.

A number of mechanisms are available to organizations for communicating change. These include:

- Team briefings
- Management meetings
- Staff meetings
- In-house broadsheets
- Company magazines
- One-to-one communication
- The use of notice boards.

Whatever method is used, communication should be swift and directed at those aspects causing stress associated with uncertainty, ignorance, fear of loss of the job, possible role ambiguity or role conflict, or even loss of face amongst a peer group. The objective is to eliminate anxiety and its associated state of tension, worry, guilt, insecurity and the constant need for reassurance.

7.12 Creating a healthy workplace

Apart from duties on employers under current health and safety legislation, such as the Workplace (Health, Safety and Welfare) Regulations, to provide and maintain a safe and healthy workplace and working environment, the provision of a healthy workplace goes a long way in the prevention or reduction of stress amongst employees.

When considering a healthy workplace, there are a number of factors that must be considered, some of which are controlled by regulations. These factors can be classified as follows.

7.12.1 Environmental factors

Environmental stressors can be classified under four headings:

- **Physical stressors**, such as stress associated with extremes of temperature, inadequate lighting and ventilation, excessive humidity, noise and vibration, radiation and inadequate workplace design.
- **Chemical stressors**, including substances hazardous to health, which may be toxic, corrosive, harmful or irritant, and in the form of gases, vapours, mists, etc.
- **Biological stressors**, including various forms of bacteria, viruses and zoonoses.
- **Work-related (ergonomic) stressors** due, for example, to repetitive movements of joints and including the risk of work-related upper limb disorders, such as tenosynovitis.

7.12.2 Cultural and social factors

It is important that employees have a sense of belonging, purpose and mission to the organization. They should have some degree of control over their work and freedom from harassment whilst at work. Factors for consideration in this case are:

- achieving a satisfactory balance between work and family commitments
- employee involvement in the decision-making process
- the option for flexitime working
- good peer communication
- training and development
- employee satisfaction with work
- positive supervisor communication and feedback
- attention to employee morale and
- ensuring an appropriate social atmosphere at work.

7.12.3 Lifestyle factors

Workplaces that support health practices encourage healthy behaviour and coping skills. These include attention to:

- smoking cessation programmes
- healthy eating and weight control
- personal hygiene measures
- physical activity on a regular basis
- women's health issues, such as ensuring a healthy pregnancy
- alcohol and drug misuse
- coping with shiftwork and
- stress management.

These arrangements and intentions should be incorporated in the organization's Statement of Health and Safety Policy under the Health and Safety at Work etc. Act 1974.

7.13 Health promotion arrangements: Organizational interventions

Employers have a duty to maintain the health of employees and to provide a workplace which is without risk to health. Preventing work-related stress is an important element of this duty and must feature in any health surveillance arrangements and in the management style.

Managers need to be knowledgeable about the causes and symptoms of stress. They need to be able to distinguish those employees who are competent, skilled and confident in their work from those who may be vulnerable to stress. In the latter case, it may be necessary to take measures to minimize stress on such persons.

Ill health resulting from stress at work should be treated in the same way as any other health hazard and should be borne in mind when assessing possible health risks in the workplace. Managers must take steps to prevent or reduce the impact of stress on employees. This takes place by certain types of 'intervention', directed at resolving the situation as far as possible.

7.13.1 Primary intervention

This proactive form of intervention, directed at preventing stress, is concerned with the work being undertaken. The purpose of primary intervention is to identify the possible causes of stress, plus the level of risk to individuals and the organization as a whole. This will be achieved through a risk assessment which should be trying to answer such questions as:

- What are the sources and levels of stress?
- How is stress affecting the health of employees?
- How is stress affecting performance in the workplace?

- How knowledgeable are the employees about managing stress?
- What additional support is needed for employees experiencing stress?

The assessment should review ways in which the situation can be improved. This may include improvements in the way employees are managed and also advice on how staff can be helped to manage their own levels of stress.

The risk assessment process can be assisted by the use of a stress audit (see later in this chapter). For this audit to be successful and beneficial, it is recommended that all employees should be encouraged to participate, the audit should be specifically designed taking into account the organization's culture and philosophy, and it should be undertaken by an external agency to guarantee confidentiality and objectivity to all participants.

The outcome of an assessment can include the redesign of tasks, establishing a better communication system between manager and employee and various supports for the employee in terms of advice and assistance from nominated employees in specific situations. Better communications between employees can be beneficial in a number of ways. One of the objectives is to get the work done in a more efficient manner.

In the case of small organizations it may be more practical to simply:

- monitor absenteeism, in particular those taking frequent spells of short-term sickness absence
- monitor staff turnover
- monitor lateness
- talk to employees on a one-to-one basis to maintain trust and confidentiality
- take note of recommendations from employees as to how things can be improved.

7.13.2 Secondary intervention

The secondary intervention level sets out to improve the overall situation at work by implementing the recommendations arising from the risk assessment. It is directed at increasing the awareness of employees to the potential for occupational stress by providing information and training whereby they are in a position to recognize and deal with work-related stress. They need to be advised of the symptoms of occupational stress at this stage.

In the first place, only employees who have the necessary skills to do the work should be recruited. Where jobs subsequently change, information and training should be provided at this stage of change and not later.

As part of the measures for reducing stress, employees should be encouraged to talk about the stress in their work and to make suggestions for reducing it.

Secondary intervention may also take place through stress management training for both managers and employees.

7.13.3 Tertiary intervention

This stage deals with the treatment and rehabilitation of those individuals who have suffered ill health as a result of stress. It may entail treatment by an occupational physician and occupational or clinical psychologist.

This stage may also entail counselling of the individual by a competent counsellor and, with the consent of the individual, the provision of feedback to the employer as to the perceived causes of stress.

7.14 Health surveillance arrangements

Health surveillance implies the on-going assessment of people's physical and mental health by a trained occupational health practitioner, such as an occupational physician or occupational health nurse. It concentrates on two main groups of employees:

- Those at risk of developing further ill health or disability by virtue of their present state of health, such as people exposed to hazardous substances and
- Those actually or potentially at risk by virtue of the type of work they undertake during their employment, such as radiation workers.

Health surveillance is one of the outcomes of a risk assessment under the Management of Health and Safety at Work Regulations. Risk assessment techniques should consider the risk of psychiatric illness arising from stress at work and make recommendations on measures for preventing or controlling this risk.

Many health surveillance procedures incorporate the personal stress questionnaire shown in Table 7.1.

7.15 Personal stress questionnaire

The personal stress questionnaire is a standard form of health surveillance directed at identifying an individual's current level of stress. It takes the form of a questionnaire shown in Table 7.1. A study of completed questionnaires should identify where there is a need for a stress risk assessment.

Table 7.1 Personal stress questionnaire

		Yes/No
	Job factors	
1.	Do you have to do too much work?	
2.	Do you have too little work to do?	
3.	Are you subject to time pressures or deadlines?	
4.	Are the physical conditions working satisfactorily?	
5.	Do you have to cope with other people's mistakes?	
6.	Do you have to make too many decisions?	
	Role in the organization	
7.	Do you suffer from role ambiguity or role conflict?	
8.	Do you have too little responsibility?	
9.	Do you participate in the decision-making process?	
10.	Are you responsible for people and things?	
11.	Do you receive adequate managerial support?	
12.	Is the organization increasing standards of acceptable performance?	
13.	Are you subject to organizational boundaries?	

Table 7.1 (*Continued*)

		Yes/No
	Relationships within the organization	
14.	Do you have good relationships with the boss?	
15.	Do you have good relationships with colleagues?	
16.	Are there difficulties in delegating responsibility?	
17.	Are there personal conflicts within the organization?	
	Career development	
18.	Do you feel you have been over-promoted?	
19.	Do you feel you have been under-promoted?	
20.	Do you suffer from lack of job security?	
21.	Do you suffer from fear of redundancy or retirement?	
22.	Are you concerned that your skills may be obsolete?	
23.	Have your ambitions been thwarted?	
24.	Do you have a feeling of being trapped in the organization?	
	Organizational structure and climate	
25.	Are there restrictions on behaviour, e.g. budgets?	
26.	Is there a lack of effective consultation and communication?	
27.	Are you uncertain as to what is happening within the organization?	
28.	Do you have a sense of belonging to the organization?	
29.	Do you feel you have lost your individual identity?	
30.	Do you suffer from office politics?	
	HOW WELL DO YOU COPE?	
	Prevention/preparation	
1.	Do you keep fit?	
2.	Do you get enough good quality sleep?	
3.	Do you eat sensibly and watch your weight?	
4.	Do you regularly take exercise?	
5.	Do you anticipate stressful events and prepare yourself for dealing with them?	
6.	Do you plan a strategy for getting through a stressful event?	
	Coping/reducing anxiety	
7.	Do you consciously try to relax?	
8.	Do you consciously breathe deeply?	
9.	Do you try to control your feelings?	
10.	Do you let your feelings show?	
11.	Do you keep cool and refuse to be rushed into anything?	
12.	Do you ask others for help?	
13.	Do you sound off to a 'safe' friend or colleague?	
14.	Do you make a decision and then stick to it?	
	Problem-solving activities	
15.	Do you seek more information about the problem?	
16.	Do you set priorities and do what you can?	
17.	Do you clarify what is expected of you?	
18.	Do you refuse to meet unreasonable demands?	
19.	Do you ask colleagues for help?	
20.	Do you ask your supervisor/boss for help?	
21.	Do you offload on colleagues?	
22.	Do you tell your supervisor/boss how you are feeling?	
23.	Do you talk about it to someone outside work?	
24.	Do you refuse to dwell on it when it is over?	

(*Continued*)

Table 7.1 (*Continued*)

		Yes/No
	Making demands on the organization	
25.	Do you ask for training on specific issues?	
26.	Do you ask for clarification about priorities and procedures?	
27.	Do you ask for more supervision?	
28.	Do you ask for time off?	
29.	Do you ask for a reduction or change in your workload?	
30.	Do you ask to change jobs?	
31.	Do you ask for formal support systems to be set up?	
32.	Do you seek out others who are feeling the same and organize for change?	
	ACTION In the next week In the next month In the next year Information, instruction and training Health surveillance Date of next review audit	

7.16 Work-related stress risk assessment

An analysis of completed personal stress questionnaires for a group of people is the starting point for the risk assessment process. Two specimen work-related stress risk assessments are shown below.

Job title: EHS administrator

		Risk
1.	Potential work stressors identified	Risk
1.1	Job design	
1.1.1	*Lack of control over pace of work* The job holders are required to respond to both internal and external customers and additionally provide support to several members of the department.	High
1.1.2	*Lone working* The job holders work in a department where occasionally they may be the only available member of staff working in the office and thus assumes responsibility for taking note of all enquiries received during the course of the working day.	High
1.1.3	*Opportunities to contribute to ideas* Whilst opportunities exist for the job holders to contribute ideas for planning and organization, as the most junior members of the department assertive behaviour may not always be demonstrated.	Low
1.2	Workload	
1.2.1	*Targets that are stretching but reasonable* Whilst flexible work schedules exist, variable customer demand may lead to several conflicting priorities.	Medium

1.3 Relationships at work

1.3.1 *Interpersonal skills*

As the most junior members of the department, the job holders occasionally need to be assertive with peers, superiors and customers. Low

1.4 Job demands

1.4.1 *Emotional demands*

The job holders deal on a daily basis with external customers most of whom are polite and reasonable. Occasionally, however, customers may become rude and aggressive. Low

1.5 Organization

1.5.1 *Change*

The company has undergone major internal reorganizational change which resulted in changes to job descriptions beyond the control of the job holders. Medium

2 Control measures required

2.1 Job design

2.1.1 *Lack of control over pace of work*

The job holders to be given clear direction that enquiries from external customers come first, then internal relations and then other members of the department. Requests for work to be done from other members of the department to be accompanied by a 'date required by' note and mutually agreed.

2.1.2 *Lone working*

Team members to make use of a movements chart on a shared drive. Other than in exceptional circumstances, a minimum of two additional members of the department to be in the office at all times. In cases where lone working is unavoidable, additional support from the Legal or Insurance Departments should be sought and agreed in advance. In addition, when it is identified that all senior members of the department will be off-site, a 'duty manager' will be nominated, who will regularly ring the office to check for messages and any problems.

2.1.3 *Opportunities to contribute to ideas*

Regular team meetings to be held with a standing agenda item to look at working practice and ideas for improvement.

2.2 Workload

2.2.1 *Targets that are stretching but reasonable*

By the nature of the company business and the job, workload from external customers is variable and not under the control of the job holders. When unreasonable/high workload demands are evident, arrangements must be made to forward excess work to other members of the department in a clear, hierarchical fashion.

2.3 Relationships at work

2.3.1 *Interpersonal skills*

Training in assertive behaviour should be provided and the job holders should be made fully aware that it is acceptable to say 'No', particularly if they are asked to do a task for which they do not feel competent.

2.4 Job demands

2.4.1 *Emotional demands*
As above, training in assertive behaviour is recommended to enable the job holders to deal with external customers who may potentially become aggressive.
2.5 Organization
2.5.1 *Change*
Annual performance appraisals to give the job holders the opportunity to discuss changes in job profiles and assess training needs as appropriate. Job descriptions must be updated.

3. Implementation plan for control measures
3.1 Job design
3.1.1 *Lack of control over pace of work*
Clarity over priority work to be emphasized at annual performance appraisals to be conducted by .... (date).
Requests for work to be accompanied with 'required by' date *with immediate effect*.
3.1.2 *Lone working*
Movements chart is available on shared drive. *With immediate effect* team members to keep electronic calendar up to date and to check movements chart to ensure sufficient cover is given in the department at all times. When team members are out of the office, an emergency contact number should be given and a nominated daily 'EHS duty manager' will regularly ring the office to check on messages.
3.1.3 *Opportunities to contribute to ideas*
Regular team meetings to be arranged (as a minimum monthly) *with immediate effect*.
3.2 Workload
3.2.1 *Targets that are stretching but reasonable*
A clear and well understood (documented) system for sharing workloads to be agreed at next EHS meeting and implemented by .... (date)
3.3 Relationships at work
3.3.1 *Interpersonal skills*
Training courses in assertive behaviour required by .... (date)
3.4 Job demands
3.4.1 *Emotional demands* – as above (3.3.1)
3.5 Organization
3.5.1 *Change*
Performance appraisals to be conducted by .... (date) and agree training needs for ..... (year). Job descriptions to be updated by ...... (date)

4. Review date: .... (date)

Commentary

This case is one which illustrates the value of undertaking stress risk assessments in a systematic way.

When the four post-holders were asked what their main job stressor was, they were unanimous that it was 'workload'. They were never able to complete all their daily tasks despite their best efforts.

However, when the risk assessment was carried out, it emerged that a major factor was the job design, especially 'lone working'. The job holders were regularly answering calls from their clients, which should have been taken by the EHS managers. The latter were very frequently off site because of the nature of their role (carrying out investigations, audits and training). The solution included a 'movements chart' kept up to date, and nominating a daily 'duty manager' who would take these calls. The administrators now had time to complete their work, the managers were happy that they were answering calls that were appropriate to them, and the clients were happier too. This was a win-win solution at no extra cost.

Job title: Clerical assistant (part-time 15 hours per week)

1. **Potential work stressors identified which may be particularly relevant in this case**

1.1 **Lack of job clarity** is identified as a low risk potential stressor. This is because the job description is seen to reflect accurately the duties and responsibilities. However, for a vulnerable person, even minor changes in duties or responsibilities may have an adverse effect even if the possible effects of any change are not carefully thought through, discussed with the post holder, and any further supportive measure, e.g. additional training, put in place.

1.2 **Job design** is identified as a low risk potential stressor in terms of *proper use of skills*, because of potential under-utilization of skills, and a medium risk potential stressor in terms of *lack of control over pace of work*. The latter is because the post holder may on occasions have competing demands from members of the ... Unit to meet deadlines.

1.3 **Workload/work schedule** is identified as a medium risk potential stressor. This is because the ... Unit's workload fluctuates on occasions, with the peaks in workload usually occurring during ...

1.4 **Relationships at work** are identified as being low risk in terms of potential to cause stress.

2. **Individual functional issues/restrictions**

.... is known to be vulnerable to stress. She has previously retired on medical grounds because of stress with a different employer.

Her specialist has warned us in the past that 'her threshold to stressful situations is very low'.

Evidence from her specialist, and from her previous employment at ... indicates that she has specific problems with time pressures and deadlines, multitasking and interruptions. These must therefore be borne in mind when assessing her particular risk from potential stressors.

The control measures considered at generic level may thus need to be augmented in this person's case. The likely additional measures will be discussed in the next section.

3. Additional control measures required

3.1 Lack of job clarity

As long as her duties and responsibilities remain clear and unchanged, risks from this area should be low. However, it is recommended that this arrangement continue to be regularly reviewed in the long term. Also it is proposed that any required training will be provided on internal training and development workshops and events. Such training as is necessary for this person should be implemented as a priority.

3.2 Job design

Under-utilization of skills has not been identified as an issue for this person before. However, her specialist has recommended that 'there should be some mechanism for monitoring her progress at work and if there are any problems these should be dealt with early before they get out of hand'. Such monitoring will need to be in place and should identify issues such as the proper use of her skills and whether they cause this person any stress.

Lack of control over pace of work will be controlled by processes that … will be put in place to ensure that she receives proper guidance should conflicting priorities occur. In view of her particular vulnerability in this area, it is critical that these processes are adequately implemented and maintained at all times, especially as the person specification form lists the 'ability to organize work and observe priorities and deadlines' as 'essential'.

3.3 Workload/work schedule

This is to be controlled by effective liaison with … to ensure a steady flow of work. For the same reasons as in 3.2, it is essential that this effective liaison to control the steady flow of work is maintained.

3.4 Relationships at work

Risks in this area are to be controlled by the Director of … being in regular contact and intervening where any conflicts or difficult situations arise. The control measures would appear adequate in this area although, in addition, in all cases there will need to be clear grievance procedures for bullying and harassment, and equal opportunities issues, etc.

4. Implementation plan for additional control measures

The implementation plan should include a monitoring system, as in 3.2. Such monitoring should also include regular review by the Occupational Health Department, and she should appreciate that it is her responsibility and in her interests to attend such reviews. The processes (as in 3.2 and 3.3) to control the flow of work and guide her through any conflicting priorities are critical for her health and safety at work. Training that is identified as being essential for this person should be implemented as a priority.

5. Review date

………… (date) or sooner if further problems arise.

Signed__________________ ________________________

Assessor Responsible manager

Commentary

This 'additional' risk assessment for a vulnerable individual was used in a pre-employment situation and allowed the managers to identify the possible areas of mismatch, or where additional control measures would be required.

In this case the manager decided not to offer her the post. It was concluded that the control measures intended to reduce fluctuations in workload, or reducing pressures from deadlines, for example, could not be sufficiently robust to ensure that her inability to cope with deadlines and interruptions would not be detrimental to her health.

However, her case worker agreed that the risk assessment had identified relevant areas of concern and advised her against taking the employer to an employment tribunal for disability discrimination. She was later employed in a more suitable post using the same risk assessment process.

7.17 Ergonomics and stress

Ergonomics is defined in several ways:

- The scientific study of work
- The study of the relationship between man, the equipment with which he works and the physical environment in which this 'man–machine system' operates and
- Human factors engineering.

It is a multidisciplinary study incorporating the expertise of a number of specialists – engineers, organization and methods study specialists, health and safety specialists, occupational hygienists, occupational health nurses and occupational physicians. As such, ergonomics seeks to create working environments in which people receive prime consideration and is an important element in the consideration of human factors at work, in particular, the potential for human error.

Under the Management of Health and Safety at Work Regulations employers must take into account the mental and physical capabilities of employees as regards health and safety when allocating tasks. This may entail recourse to ergonomic principles, with particular reference to psychological factors, such as perception, memory, attitudes, vigilance, information processing, learning and individual skills. Physical factors, such as strength, stamina and certain body dimensions, such as height and arm span, may also need to be considered, together with environmental factors and the potential for environmentally related stress.

7.17.1 Elements of ergonomic study

Fundamentally, ergonomics covers four main areas – the total working system. These areas are significant in the design of working layouts, the setting of work rates, the arrangement of safe systems of work and in the prevention of stress. The total working system is outlined in Table 7.2.

Table 7.2 The total working system

Human characteristics	Environmental factors
Body dimensions Strength Physical and mental limitations Stamina Learning Perception Reaction	Temperature Humidity Light Ventilation Noise Vibration
Man–machine interface	**Total working system**
Controls Communications Automation	Work rate Posture Stress Productivity Accidents Safety

The human system

This area is concerned with the principal characteristics of people in terms of the physical elements of the body dimensions, strength and stamina, together with the psychological elements of learning, perception and reaction to given situations. It is further concerned with the allocation of tasks to people based on the individual physical and mental capabilities. Failure to take these factors into account in the design and allocation of tasks can result in stress on employees.

Environmental factors

Environmental factors, such as noise, temperature and humidity have a direct effect on the performance of employees at work. Poor environments cause stress on operators, resulting in poor standards of performance.

The man–machine interface

Machinery incorporates:

- Displays, which provide information to the operator
- Controls, which may be of the manual, mechanical or electrical type
- A number of design features specific to each machine.

This man–machine interface should be designed with a view to eliminating operator error, a significant cause of stress. In many cases, machinery is designed with the principal objective of maximizing production with little consideration for the operator in terms of ease of operation, comfort whilst working and safe operation.

Total working system

This fourth element summarizes the outcome of failure to consider the other three elements in the design of the man–machine interface. Poor design results in operator fatigue, inadequate work rates and, in some cases, the need to adopt awkward postures while operating machinery. The outcome can be reduced productivity, excessive stress on the operator, who may be pressurized to meet production targets, an increased potential for accidents and, generally, reduced standards of safety.

7.17.2 Design elements

Factors such as the location, reliability, ease of operation and distinction of controls, together with the identification, ease of reading, sufficiency, meaning and compatibility of displays are all important in ensuring safe and correct operation of machinery of all types.

7.18 Job design and organization

7.18.1 Features of jobs

All jobs incorporate a number of specific characteristics and demands on operators which, in most cases, require that they be provided with information, instruction and training prior to commencing the job. Moreover, all work situations incorporate a series of 'socio-technical factors' associated with, for instance, the communication system within the group, working hours and manning levels. Organizational characteristics of jobs, such as the allocation of individual responsibilities, also feature in the design of jobs.

The various characteristics of jobs are shown below:

Job characteristics

All jobs, no matter how simple they may be, incorporate a number of characteristics, including:

- The frequency of operation of, for instance, a vertical drilling machine
- The repetitiveness of the job
- The actual workload
- The criticality of the job in terms of its accurate completion according to a prescribed procedure
- The duration of the job
- The job's interaction with other jobs as part of a work process.

Job demands

The characteristics of a job impose various physical and mental demands on the worker. Some tasks may require a high degree of physical strength and stamina, such as manual handling activities. Other tasks, such as those involving inspection of finished products, need a high degree of attention and vigilance in ensuring products meet their specification, with rejects being removed and set aside for further attention.

Instructions and procedures

Any instructions and information must be comprehensible to the operator and relevant to the job in hand. The quality of both verbal and written instructions to operators and the formal procedures established have a direct relationship with the potential for human error. All instructions and procedures, whether orally or in writing, should be clear, comprehensible, relevant, unambiguous, sufficient in detail, easy to use, accurate and produced in an acceptable format. Instructions and procedures should be subject to regular revision, particularly as a result of complaints by operators where they may be experiencing difficulties in interpretation and use.

Job stressors

Many jobs, through the failure by the employer to consider the physical and mental needs of the operator, his level of intelligence, attitude and motivation, can be stressful. The following questions should be asked at the design stage of jobs and at the job analysis stage.

- Does the task isolate the operator, both audibly and visually, from other operators?
- Does the task put operators under pressure due to the need to complete the task within a prescribed timescale?
- Does the job impose a higher level of mental and physical workload on people than normal?
- Is the job of a highly repetitive nature?
- Do the various jobs create conflict amongst employees?
- Do some tasks result in physical pain or discomfort, such as manual handling operations or work at low temperatures?
- Is there a risk of distraction in some critical tasks?
- Is there sufficient space available to undertake the work safely?
- Where a shift work system is in operation, does this system take account of the physical and mental limitations of employees?
- Where an incentive scheme is in operation, are the incentives offered seen as fair to all concerned?

The socio-technical factors

Socio-technical factors cover a wide range of issues which can contribute to stress. These include:

- The social relationship between operators and how they work together as a group
- Group working practices
- Working hours
- Provision of meal and rest breaks
- Formal and informal communication systems
- Rewards and benefits available.

Organizational features also come within this area of consideration. These include:

- The actual structure of the organization and the individual work groups
- Allocation of responsibilities

- Identification of authority for certain actions
- Interface between different work groups.

7.18.2 Design ergonomics

This is an area of ergonomic study concerned with the design and specification of the various elements of the man–machine interface, in particular controls and displays to various forms of work equipment, directed at reducing stress on operators.

Controls include various forms of physical controls to machinery, plant and vehicles, such as levers, brake levers, steering wheels, switches and control buttons, foot pedals and gear sticks. Displays, on the other hand, provide visual information to the machine operator or driver and include speedometers, fuel gauges, pressure gauges and clocks.

The principal design features of the man–machine interface are outlined below.

Layout

The layout of working areas, workstations and operating positions should allow for free movements, safe access and egress, and unhindered visual and oral communication between operators. Congested, badly planned layouts, as seen with some supermarket checkout stations, for example, result in stress and operator fatigue and increase the potential for accidents.

Vision

The operator should be able to set and use controls and read dials and instruments with ease. This reduces fatigue and accidents arising from faulty or incorrect perception.

Posture

The more abnormal the working posture, the greater the potential for postural fatigue and long-term injury. Work processes and systems should be designed to permit a comfortable posture which reduces excessive job movements. Not only must this be considered in the siting of controls, but also basic requirements, such as the correct height of workbenches which prevent stooping during assembly or inspection tasks, for example.

Comfort

The comfort of the operator, whether driving a vehicle or operating machinery, is essential for his physical and mental well-being. Environmental factors, such as temperature and ventilation, directly affect comfort and should be given priority.

7.19 Stress management programmes

Many organizations run stress management programmes for employees. These programmes have a range of objectives, including:

- To define 'stress', to classify the stressors at work and to promote a greater understanding of the causes of stress

- To train delegates to recognize the causes and effects of stress and their individual stress responses
- To provide a means for delegates to measure and evaluate stress
- To consider stress reduction strategies at individual level and in the workplace
- To consider strategies for coping with the stress in our lives.

It should be recognized that a stress management programme is a continuing process. It should be run by trained occupational health practitioners who not only provide guidance to employees on the subject but provide a range of services, such as counselling, hypnotherapy and other stress-reducing techniques.

7.20 Occupational health schemes and services

Occupational health is a branch of preventive medicine concerned, firstly, with the relationship of work to health and, secondly, the effects of work on the worker. With the increasing attention to health at work, many employers have installed in-house occupational health services or provided access to occupational health schemes for employees. Staffed by occupational physicians, occupational health nurses, occupational psychologists and occupational hygienists, such services incorporate the principal areas of occupational health practice, thus:

- Placing people in suitable work
- Health surveillance
- Providing a treatment service
- Primary and secondary health monitoring
- Advice on avoiding potential risks to health
- Monitoring for the early evidence of non-occupational disease
- Counselling
- Health education
- First aid and emergency services
- Welfare amenity provision
- Environmental control and occupational hygiene
- Liaison with enforcement officers, medical and nursing advisers of the Employment Medical Advisory Service and general medical practitioners and
- The maintenance of health records.

Advice is commonly provided on a range of other matters, such as hearing and eyesight defects, drug addiction, smoking and alcoholism.

7.21 Remedies for employers

Employers need to accept that stress is a potential risk and that employees may experience stress at work. This must be accepted at executive and senior management level if any stress management programme is to succeed. Broadly, they need to consider and put into practice a corporate strategy for dealing with stress at work. Such a strategy incorporates four principal elements:

7.21.1 Recognition/identification of workplace stressors

Recognition of stress can be achieved by simply talking to employees with a view to identifying areas of dissatisfaction and disillusionment with their work. A study of current sickness absence data may also identify trends, particularly with respect to people taking regular periods of short-term sickness absence. Evidence of regular lateness for work, staff turnover and certain disciplinary issues may also indicate sources of stress. However, many people are reluctant to admit to stress arising from work through fear of losing their job, being 'sidelined' into another less responsible job, fear of ridicule that they no longer have 'what it takes' to do the job or personal guilt at manifesting signs of stress.

Sources of information on workplace stressors are:

- Staff complaints
- Job descriptions
- Work-related ill health and sickness absence data
- Performance appraisals/job and career reviews
- Reports from occupational health practitioners, such as occupational health nurses
- Overtime reports
- Accident investigations.

7.21.2 Measurement and evaluation/assessment of stress

A number of techniques and procedures are available aimed at assessing the level of stress amongst employees. They include a number of confidential techniques involving counselling by trained specialists such as occupational psychologists and occupational health nurses, the use of health questionnaires and personal stress audits, and health screening operations by occupational health practitioners.

Conducting an anonymous survey using a well-designed questionnaire will give an indication of the stress currently existing within the organization.

It should be appreciated that the risk assessment provisions of the Management of Health and Safety at Work Regulations can apply in cases of work-related stress and that it may be necessary to undertake a formal risk assessment to assess current stress levels in the workplace.

As with any risk assessment, a record must be kept of the significant findings arising from the risk assessment, namely:

- A record of the preventive and protective measures in place to control the risks
- What further action, if any, needs to be taken to reduce risk sufficiently
- Proof that a suitable and sufficient assessment has been made.

Typical stressors that might be encountered in a risk assessment of this type and the management actions necessary are indicated below in Table 7.3.

Table 7.3 The stress elements of risk assessment

Stressor	Management action
Lack of control over work operations	Provide opportunities for employees to contribute ideas, perhaps through a suggestion scheme
Poor working relationships amongst individuals in a work group	Provide training in interpersonal skills
Inflexible work schedules	Examine the scope for flexible work schedules and implement same if possible
Repetitive, insufficient or boring work	Change the way jobs are undertaken by job rotation, allocating individual responsibilities, extending the scope of jobs, increasing the variety of tasks, greater responsibility for the performance of the group
Confusion as to who does what	Ensure everyone has clearly identified objectives and responsibilities which are linked to the organization's business aims
Overload situations – too much to do in the time available	Advise employees well ahead of any urgent jobs which will require a greater effort, prioritize tasks and eliminate unnecessary work
Bullying, racial and sexual harassment	Produce and put into operation a formal policy on these issues, including disciplinary procedures. Establish procedures, including confidential reporting arrangements, investigation procedure and agreed grievance procedure
Lack of communication and consultation	Introduce clear business objectives, formal and informal communication arrangements, together with close employee involvement particularly prior to and during change
The need to meet formal training requirements for the job	Ensure people are matched to their jobs before requiring them to undertake the training; ensure they have the mental capacity to receive the training; there is nothing worse than failing a training course because of poor selection
Having responsibility for looking after others	Provide training and support, particularly where tasks may incorporate an element of risk
Physical dangers arising from, for example, electricity, poor working conditions, noise, hazardous substances and contact with members of the public	Ensure adequate prevention and control strategies are in place prior to work commencing

Any risk assessment must be subject to regular review and, where more than five employees are employed, the significant findings of the risk assessment must be recorded. Monitoring and evaluation are essential elements for the development of effective stress management programmes. The Holmes–Rahe Scale of Life Change Events is commonly used for evaluating personal stress levels.

7.21.3 Strategies for eliminating, reducing or controlling stress

A range of strategies are available including:

- Changes in work practices
- Relocation of individuals
- Provision of additional training, such as stress awareness training
- Stress therapy, such as Progressive Muscular Relaxation techniques
- Job rotation
- Redesign of working practices, such as flexible shift work

- Modified behaviour on the part of the individual
- Reduction in working hours
- Change in responsibilities
- Increased participation in the decision-making processes for jobs.

These strategies, linked with a corporate Stress Management Programme, can be very effective in reducing stress amongst employees. They require a concerted co-ordination exercise between human relations managers, senior managers, line managers and occupational health practitioners in order to be effective.

7.21.4 Monitoring and review

It is essential that the individual's health is monitored by an occupational health practitioner on an on-going basis to ensure that his ability to cope has improved as a result of the measures taken and that there has been no deterioration in terms of reversion to stress-related behaviour and subsequent ill health. This may take the form of monthly health consultations with an occupational health nurse, supported by various stress-relieving techniques and therapies.

7.22 FIET recommendations on limitations of work-related stress and pressure affecting salaried employees

In 1992 the International Federation of Commercial, Clerical, Professional and Technical Employees (FIET) made the following points with respect to formulating a policy on occupational stress:

1. Preventive health protection starts with the way the work is organized.
2. Trade union representatives must be involved in a comprehensive and timely fashion in deciding how work is organized and carried out.
3. This involvement also includes staffing requirements (personnel planning), as well as the introduction or modification of personnel information and performance evaluation systems.
4. Work schedules which cause great strain (e.g. night and shift work) are to be eliminated or reduced.
5. Work is to be organized in such a way that the individual worker can have independence and responsibility.
6. Measures to prevent, alleviate or compensate work-related pressures must not be subordinate to purely economic considerations.
7. Initial and further training opportunities must be offered which take current and future qualification requirements of employees into consideration. This also includes a role for the trade unions in determining the contents of training courses in order to ensure that preventive health protection is included.

8. Individuals' rights to more self and co-determination at and about work must be revised.
9. Legislators are called upon to take into account the increase of stress-related illnesses by further developing social legislation (recognition of stress-related illnesses as occupational illnesses).
10. As an accompaniment to the above-named measures, environmental measures at the workplace must be further developed in collective and company agreements.

7.23 Termination of employment for work-related stress

In a situation where an employee complains to his employer of work-related stress, the employer may take the unwise action of terminating that person's employment. This, of course, exposes the employer to the possibility of a civil action by the employee for breach of contract (see Chapter 8).

An alternative course of action for the employee is an action for unfair dismissal or discrimination before an employment tribunal. The jurisdiction of employment tribunals is based entirely on statute, in particular the Employment Rights Act (ERA) 1996. This Act stipulates that every employee has the right not to be unfairly dismissed by the employer. To make an application for unfair dismissal to an employment tribunal, an employee must have continuous employment with the same employer for at least 1 year prior to the effective date of termination of employment, and must have been 'dismissed'.

Dismissal may occur at both common law and by statute. However, common law only recognizes one act which constitutes dismissal, that is the termination of the contract by the employer with or without notice. On the other hand, the ERA recognizes two further acts which constitute dismissal, namely the non-renewal of a temporary contract, and 'constructive dismissal' where, for example, an employee becomes so stressed by some aspect of his work or the working environment, that he feels obliged to resign.

Under the ERA, an employee shall be treated as dismissed by an employer, if:

- The contract under which he is employed is terminated by the employer, with or without notice or
- The employee is employed under a contract for a fixed term, that term expires without being renewed under the same contract or
- The employee terminates the contract under which he is employed, without notice in circumstances in which he is entitled to terminate it without notice by reason of the employer's conduct.

It is the last of these circumstances which is termed 'constructive dismissal'. However, in an action for constructive dismissal, it is for the employee to show that the employer had committed a fundamental breach of the contract of employment.

Under section 98 of the ERA, it is for the employer to show the reason for dismissal and that it is a reason which relates to capability qualifications, conduct, redundancy, contravention of a statute or some other substantial reason of a kind such as to justify the dismissal of an employee holding the position which the employee held. ('Capability', in relation to an employee, means his capability assessed by reference to skill, aptitude, health or any other physical or mental quality.) On this basis, stress or stress-related ill health comes within this area of capability.

This section goes on to state 'the determination of the question whether dismissal was fair or unfair, having regard to the reason shown by the employer, depends on whether in the circumstances, including the size and administrative resources of the employer's undertaking, the employer acted reasonably or unreasonably in treating it as a sufficient reason for dismissing the employee, and that question shall be determined in accordance with equity and the substantial merits of the case' (see further Chapter 9).

7.24 Stress management action plans

Organizational level

Any action plan to deal with stress at the organizational level should follow a number of clearly defined stages, as follows:

1. Recognize the causes and symptoms of stress.
2. Decide the organization needs to do something about it.
3. Decide which are the group or groups of people in whom we can least afford stress, e.g. key operators, supervisors.
4. Examine and evaluate by interview and/or questionnaire the specific causes of stress.
5. Analyse the problem areas.
6. Decide on suitable strategies, e.g. counselling, social support, training, such as time management, environmental improvement and control, redesign of jobs, ergonomic studies.

In some cases, employees may need some form of rehabilitation where they demonstrate measurable stress and/or mental health symptoms. Employee Assistance Programmes are commonly used to provide help at this level, including the detection of problems and prompt referral of affected individuals for specialist treatment. The vast majority of those who develop mental illness make a complete recovery and ultimately return to work. What should be appreciated is that it is far more costly to retire an employee on medical grounds and recruit and train a successor than it is to spend time and effort ensuring satisfactory rehabilitation of an existing employee.

Individual level

At the individual level, people should take the following action:

1. Identify your work and life objectives. Re-evaluate on a regular basis or as necessary. Put them up where you can see them.
2. Ensure a correct time balance.
3. Identify your stress indicators. Plan how you can eliminate these sources of stress. See them as red STOP lights.
4. Allow 30 min each day for refreshing and recharging.
5. Identify crisis areas. Plan contingencies.

6. Identify key tasks and priorities. Do the *important*, not necessarily the *urgent*.
7. Keep your eyes on your objectives. Above all, have fun!

7.24.1 The benefits to the organization

The benefits to an organization of having well-developed strategies for dealing with stress at work can be summarized thus:

- A contented workforce working in harmony
- Reduced sickness absence costs
- Increased performance and output
- Better working relationships between line managers and employees
- Lower labour turnover
- Better communications between individuals and departments.

7.24.2 A corporate strategy for dealing with stress at work

The Chartered Institute of Personnel and Development (CIPD) outline four main approaches that organizations can adopt to address stress at work. These can be used together as a single initiative or may be adopted individually in a more step-by-step well-being programme.

- **Policy, procedures and systems audit**: This approach requires the organization to undertake an audit of its policies, procedures and systems to ensure that it provides a working environment that protects the well-being of the workforce and is able to identify troubled employees and provide them with an appropriate level of support.
- **Problem-centred approach**: This approach provides a problem-solving model for dealing with stress and other psychosocial issues. It takes issues and problems that arise within the workplace and identifies why they have occurred and then finds ways to solve them. The identification process may involve undertaking a risk assessment, examining sickness absence levels, employee feedback, claims for compensation and performance deficits.
- **Well-being approach**: This approach takes the view that the aim is to maximize employee well-being. Although it uses similar tools to those used by the problem-centred approach it is much more proactive in identifying ways to create a healthy workforce.
- **Employee-centred approach**: This approach works at the individual level of the employee. Individuals are provided with education and support in order to help them deal with the problems they face in the workplace. The employee-centred approach focuses on employee counselling and stress management training.

Measures to reduce workplace stress

- Undertaking a stress audit and subsequently directing resources to reduce or eliminate the sources of stress

- Agenda items should include terms and conditions of employment, physical and psychological working conditions, work content, communication systems and working relationships
- The development of a supportive work ethos to encourage staff to discuss and seek support when experiencing stress.

When sources of stress cannot be eliminated, the following intervention may be considered:

- Stress management and relaxation technique training
- Promotion of healthy behaviour and exercise
- Personal counselling schemes.

Should an organization have a stress policy?

It has been found that a well-being policy is much more effective in recognizing the need to maximize the well-being of their employees rather than merely reduce their level of stress. Whether organizations choose a 'well-being' or 'stress' policy, the elements that should be contained in the policy are very similar. There is a need for a clear statement supported by senior management, which shows that the organization is committed to developing a working environment that promotes the health and well-being of the organization and its employees. The statement should be supported by a number of principles including:

- A constant review of company policies, procedures and initiatives to ensure that they maximize employee well-being
- The identification and regular review of the key well-being indicators
- The provision of effective advice, support, counselling and training to enhance employee well-being
- The process for evaluating the effectiveness of all well-being initiatives.

Stress and the employee

Excessive pressure and problems can occur in an employees' personal life as well as at work. If an employee is facing a relationship breakdown, financial problems, difficulties with neighbours or bereavement, it is possible that these problems may affect the employee's ability to work effectively. Employees do not have to describe the nature of their personal problems to their manager or personnel. However, if personal problems begin to affect performance at work then this will need to be raised and discussed with the employee. The objective of the meeting would be to identify the help that the manager or personnel can give to the employee to return to effective working.

There are a number of possible causes of change in the employee's performance. These include:

Organizational problems	*Physical problems*
a training need	physical illness
a relationship problem	design of workstation
workload or pace	noise, lighting
loss of motivation	violent attack at work

Psychological problems
anxiety or depression
phobias or panic attacks
anger management
addictive behaviours, e.g. alcoholism, gambling

Social problems
housing problems
relationship difficulties
financial problems
legal problems, e.g. divorce, custody or crime

It is important that personnel professionals are aware of what they can and cannot do when supporting employees suffering from stress. While the organizational problems need organizational solutions, some of the other problems need the support or attention of qualified experts. It can be very easy to become over involved in helping troubled employees and, therefore, it is important to refer employees for expert help when this is more appropriate.

Dos and don'ts for the personnel professional

Do	Don't
Make yourself available for employees who feel unable to talk to their manager.	Try to solve employees' problems for them.
Point the employee towards other sources of help, e.g. Citizens Advice, a counselling service, helpful books.	Try to be a counsellor without the specialist training. Leave it to the professional.
Facilitate discussions between the employee and his manager.	Act as a go-between for the employee and the manager.
Ensure that managers recognize their responsibility for protecting the well-being of their teams.	Take on the manager's responsibility for them.
Look after your own health and well-being by balancing your life.	Ignore your own needs. When you need help ask for it.

Signs of stress

The first signs that indicate employees may be suffering from excessive pressure or stress are changes in their behaviour or appearance. A guide on the kinds of changes that may occur is given below.

Work performance
declining or inconsistent performance
uncharacteristic errors
loss of control over work
loss of motivation and commitment
indecision
lapses in memory
increased time at work
lack of holiday planning/usage

Regression
crying
arguments
undue sensitivity
irritability and moodiness
over-reaction to problems
personality clashes
sulking
immature behaviour

Withdrawal
arriving late to work
leaving early

Aggressive behaviour
malicious gossip
criticism of others

extended lunches
absenteeism
resigned attitude
reduced social contact
elusiveness or evasiveness

vandalism
shouting
bullying or harassment
poor employee relations
temper outbursts

Other behaviours
out of character behaviour
difficulty in relaxing
increased consumption of alcohol
increased smoking
lack of interest in appearance and hygiene
accidents at home or work
reckless driving
unnecessary risk taking

Physical signs
nervous stumbling speech
sweating
tiredness/lethargy
upset stomach/flatulence
tension headaches
hand tremor
rapid weight gain or loss
constantly feeling cold

CIPD viewpoint

Stress in the workplace has to be properly managed if it is to be controlled. CIPD believes that:

- people work more effectively within a participation management style;
- people are better motivated when work satisfies economic, social and psychological needs;
- motivation improves by paying attention to job design and work organization.

7.25 A corporate fitness programme

With a view to improving individual performance, providing increased benefits to their employees and to reduce insurance-related costs, many organizations are developing corporate fitness and wellness programmes. The first step is to establish the objectives the organization hopes to achieve by offering such a programme.

It is easier for organizations to focus on employee morale and to improving their image in the community rather than reducing insurance costs largely because employees who would benefit from these programmes are not necessarily the people who would have taken advantage of them. However, employees who do use them are the ones who would seek out the services on their own if they were not offered by the organization.

Once the objectives have been established, it would be appropriate to review data on health insurance, employee compensation, absenteeism and baseline data, such as age, sex and job descriptions. For example, a company that has a workforce with a large female population in the age range of 20–30 years may want to focus on offering maternity and child care services. Organizations whose work entails manual handling of goods, such as distribution companies, may seek to provide training programmes for employees directed at correct manual handling techniques with a view to reducing manual handling injuries, such as muscle and ligamental strains, hernias and prolapsed intervertebral discs.

Offering health interviews by a trained professional, such as occupational health nurse, to show employees where they need to focus their efforts in improving their own well-being and fitness for work, is an important feature of a corporate fitness programme.

A number of matters need resolving prior to the introduction of such a programme.

- What are the objectives of the programme?
- What are the financial arrangements in terms of a budget for the programme?
- Will employees be permitted to participate in the programme during working hours?
- Will the facilities be provided on the organization's premises?
- What is the best way to deliver fitness services to employees?
- Who will run the fitness programme?
- What are the feelings of employees about undertaking such a programme?
- Does the insurance company offer such a programme as part of the overall insurance cover package?
- How is the programme to be promoted?
- What needs to be done to maintain employee commitment to the programme?

There should be a corporate fitness policy which states the objectives of the policy, the organization and arrangements for implementing the policy, including the individual responsibilities of directors and senior managers in terms of promoting and supporting the programme. The policy should be reviewed on a regular basis with a view to assessing whether the original objectives are being met, particularly with regard to a reduction in ill health-related lost time and improvements in morale and general fitness of the workforce.

7.26 EU principles of stress prevention

The EU lays down the following principles for designing a stress prevention strategy at the workplace:

- **Prevention through improved design**: Action must start at the design stage. Facilities, equipment, machinery and tools have to be planned having in mind the potential health impact. Company buying policies need to consider their future impact on employees' well-being.
- **Participation of end users**: Technocratic approaches should be avoided. Joint initiatives of managers, workers and professionals are the key to success, and top-down management strategies should be replaced by an involvement approach.
- **Better work organization**: This means greater control and autonomy by workers over their tasks, less monotonous and repetitive jobs with greater social interaction and work group support.
- **A holistic approach to the environment**: Both the physical and social environments should be integrated, while both the working and living environments should also be considered as one in terms of their effects on occupational stress.
- **An enabling organizational culture**: A healthy company should be measured by the quality of working life for its workforce and not only by economic values. The

benefits from such an approach should be gauged in the medium and long term, not in the short term.

- **Attention to workers with special needs**: These include workers such as shift workers, migrant workers, older and younger workers. The gender dimension also needs to be taken into account.
- **Economic feasibility**: Economic feasibility of the prevention strategy will provide support for the policies and will increase the chance of success and commitment by the organization (EC Health and Safety Directorate, 1992).

7.27 Conclusion

Some managers are not prepared to recognize the problem of stress in the workplace. However, it was not until people, such as occupational health nurses, started to relate sickness absence levels to stress that managers eventually began to admit that the results of their decisions and actions, the environment they provided for operators and many other features of their organizational activities could be stressful.

The problem is still with us, however. Many people simply fail to recognize actual stressful situations or future stressful situations in their lives. These situations can arise from problems at home, in their relationships with people or as a result of a specific life event, such as a bereavement. Generally, they 'bear up' and endeavour to cope. In most cases, they do cope but with varying effects on their health, some of which can be serious.

If people are to cope with stressful situations, they have to return to the basic principles, namely:

- identify the sorts of events in their lives which create the stress response
- measure and evaluate the significance of these events and
- learn various forms of coping strategies to enable them to deal with these life events.

People would be much happier if they were to undertake this exercise.

Questions to ask yourself after reading this chapter

- Does the organization have documented procedures for managing stress?
- Has consideration been given to the HSE Management Standards for Stress?
- Do managers frequently have to make decisions whilst under stress?
- Is the organization's system for communicating change adequate?
- What efforts are made to ensure a healthy workplace?
- Are ergonomic factors taken into account in the design of jobs?
- Does the organization run stress management programmes for employees?
- Does the organization use an occupational health service?
- Is there a corporate strategy for dealing with stress at work?

- Is there a formal Statement of Policy on Stress at Work in the organization?
- Does the organization operate a corporate fitness programme for employees?

Key points – implications for employers

- Employers need to recognize the existence of stress in the workplace and install a series of actions for dealing with it.
- Existing management systems may need reviewing and revising with a view to preventing stress.
- Many of the causes of human failure are associated with stress.
- There are considerable benefits to be achieved through eliminating stress at work, in particular better health of employees, reduced sickness absence, increased performance and output, better working relationships and lower staff turnover.
- The 'setting' or framework within which work takes place is important in preventing stress.
- Good systems of communication, particularly those directed at informing employees about future changes in the workplace, in working practices and procedures, are essential.
- Various forms of health surveillance, including counselling on stress-related issues, may be necessary to combat stress amongst employees.
- Ergonomic principles should be applied in the design of systems of work and operator workstations.
- The running of stress management programmes may be beneficial in some cases, together with the use of an occupational health service.
- Organizations need to have a formal procedure for dealing with stress-related termination of employment.

8

The civil implications

Employers need to recognize the civil liability implications of stress at work. In the last decade the number of stress at work claims in the civil courts has increased dramatically. These claims tend to attract considerable publicity with out-of-court settlements being splashed across newspaper front pages. Trade unions have also taken this issue on board and it is conceivable that the number of claims will continue to rise. Statistics produced by the Trades Union Council indicate that there were 6400 trade union-backed claims in the year 2000, compared with 519 similar claims in 1999. On the other hand, many claims have failed due to the difficulties in resolving the issues of 'foreseeability' and 'causation' outlined below.

Fundamentally, in the case of work-related stress, an employee has two choices. Firstly, he can sue his employer for breach of contract, negligence or breach of statutory duty in the County or High Court. Alternatively, he may make an application to an employment tribunal for unfair dismissal, constructive dismissal or discrimination.

8.1 The landmark case

The High Court decision in Walker v Northumberland County Council (judgement delivered by Colman on 16 November 1994) in which a senior social worker was given leave to claim up to £200 000 damages against his employers, Northumberland County Council, for allowing him to work to the point of breakdown, was the first case to focus attention on the question of stress in the workplace. It raised the question of unreasonable behaviour by employers in pushing people too far and for failing to provide effective care and the resources needed to do the job.

The employee in question, John Walker, worked for Northumberland County Council as an Area Social Services Officer from 1970 to December 1987, reporting to the assistant director of social services. As a manager, he was responsible for the day-to-day activities of various groups of field workers in his department.

During the 1980s the number of child abuse cases being handled by his colleagues in the department and himself increased dramatically. In 1986, Walker suffered a nervous breakdown which was attributed to mental exhaustion. This resulted in acute anxiety, headaches, insomnia and an inability to cope with any form of stress. He was prone to tears and was upset easily. As a result of his state of health, he took some time off from work to recuperate. When he eventually returned to work he asked his senior manager to be relieved of some of the pressure at work. As a result, the council provided a principal field worker to assist him on a temporary basis.

This assistance tended to be short-lived, however, Mr Walker being told that the principal field worker was also required to cover for other members of staff. Furthermore, as a result of Mr Walker's absence due to his first nervous breakdown, a large backlog of paperwork had accumulated which no one had attended to during this period of absence. He was requested to clear this backlog in addition to recommencing his routine duties.

After a short holiday, he returned to work but his stress symptoms increased and he informed his senior manager that he could not be expected to continue in the present situation. He went on sick leave again and was diagnosed as suffering from stress-related anxiety. During this period he suffered a second nervous breakdown and was obliged to retire from work. In 1988 he was dismissed on the grounds of permanent ill health.

8.1.1 The claim

As a result of this series of events, Mr Walker claimed damages for breach of the duty of care owed to him by the county council in that the council failed to take reasonable steps to avoid exposing him to an excessive workload which was a risk to his health. He was of the opinion that the council should have recognized that the current workload he was under could affect his health, in view of the previous indications he had given to his superiors of this excessive workload, and bearing in mind the generally stressful nature of his work dealing with child abuse cases.

8.1.2 The decision

In the view of Colman, one of the principal issues in this case was whether the first nervous breakdown was attributable to a breach of the council's duty of care. Generally, an employer has a duty to provide his employees with a safe system of work and to protect them from risks of a reasonably foreseeable kind. Whilst the law had limited this legal concept to physical injury, the duty of care could be extended to cover mental injury associated with stress at work.

A number of questions need consideration:

- At what point is it the duty of the employer to take steps to protect his employee, given that any form of professional work is intrinsically demanding and stressful?

- What assumption is the employer entitled to make about the resilience, mental toughness and stability of character of an individual, given that people of clinically normal personality may have a widely differing ability to cope with stress attributable to their work?

Colman stated that:

> *Once a duty of care has been established, the standard of care required for the performance of that duty must be measured against the yardstick of reasonable conduct on the part of the person who owes the duty. The law does not impose on such person the duty of an insurer against all injury or damage caused by him or her, however likely or unexpected.*

Fundamentally,

- the practicability of remedial measures should take into account the resources and facilities at the disposal of the person or institution owing a duty of care (*British Railways Board v Herrington (1972) AC 877I*) and
- consideration must also be given to the purpose of the activity which has given rise to the risk of injury. (In the words of Denning in *Watt v Hertfordshire County Council (1954) 1 WLR 835*, 'the risk must be balanced against the end to be achieved'.)

8.1.3 The outcome

The Walker decision raises a number of issues:

- An employer may be in breach of an implied term in failing to take reasonable care of the employee's health and safety at the workplace, which includes mental injury. This may be considered to be a fundamental breach of contract which goes to the root of the contract of employment.
- The employee may be entitled to treat the contract as having been repudiated by the employer because of the employer's unreasonable conduct in taking steps to minimize stress at the workplace, and also by exposing the employee to high levels of stress through a heavy workload. The employee could claim constructive dismissal and claim for damages.
- Under section 55(2)(c) of the Employment Protection (Consolidation) Act 1978, an employee can terminate the contract of employment 'with or without notice in circumstances such that he is entitled to terminate it without notice by reason of the employer's conduct'.
- The employer's conduct must be sufficiently serious to entitle the employee to leave at once. If the employee continues to accept the heavy workload demands and stressful situation after making the complaint, the employee may lose the right to treat himself as discharged, and will be regarded as having elected to affirm the contract.
- The onus is on the employee to notify his employer of the heavy workload demands upon him, and to identify what actions the employer proposes to take to ease the workload.

In the light of the Walker decision, some consideration should be given by an employer to identifying those factors which could result in stress at work, and to find ways of minimizing the stress levels.

8.2 Principal areas of consideration

8.2.1 The duty to take reasonable care

The position at common law is that employers must take reasonable care to protect their employees from the risks of foreseeable injury, disease or death at work.

The case of *Donoghue v Stevenson (1932) AC 562* established the 'neighbour principle', the duty owed by a person to his 'neighbours'. Here the claimant, Mrs Donoghue, purchased a bottle of ginger beer that had been manufactured by the defendant, Mr Stevenson. The bottle was of dark glass so that its contents could not be seen before they were poured out. The bottle was also sealed so that it could not have been tampered with until it reached the ultimate consumer. When the claimant poured out the ginger beer, the remains of a decomposed snail was seen floating in the glass. Not unnaturally, Mrs Donoghue was 'taken ill', having already consumed part of the contents of the bottle.

Lord Atkin, in his judgement, made the following point:

> *A manufacturer of products, which he sells in such a form as to show that he intends to reach the ultimate consumer in the form in which they left him with no reasonable possibility of immediate inspection, and with the knowledge that the absence of reasonable care in the preparation or putting up of the products will result in an injury to the consumer's life or property, owes a duty to the consumer to take reasonable care.*

Lord Atkin's judgement went further:

> *You must take reasonable care to avoid acts or omissions which you reasonably foresee would be likely to injure your neighbour, i.e. persons who are so closely and directly affected by my act that I ought reasonably to have them in contemplation as being so affected when I am directing my mind to the acts or omissions which are called in question.*

Thus, in the application of the 'neighbour principle', there must be a close and direct relationship between the claimant and the defendant, such as that between an employer and employee, manufacturer of products and consumer of same or an occupier of premises and a visitor, and secondly, that the defendant must be able to foresee a real risk of injury to the claimant if they, the defendant, do not conduct their operations or manage their property with due care.

The effect of this requirement is that if an employer knows of a health and safety risk to employees, or ought, in the light of knowledge current at that time, to have known of the risks, he will be liable if an employee is injured or killed or suffers illness as a result of the risks, or if the employer failed to take reasonable care to avoid this happening. In the case of stress at work, the civil implications are now well established, and employers cannot disregard the need to consider the potential for stress amongst employees and the measures necessary to avoid same as far as possible.

8.2.2 The employer's common law duty of care

At common law it has long been established by precedent that an employer owes a duty of care to each of his employees while they are in the course of their employment. An employer's duties under common law were identified in general terms by the House of Lords in *Wilson's & Clyde Coal Co. Ltd v English (1938) AC 57 2 AER 628.* The common law requires that all employers provide and maintain the following:

- A safe place of work with safe means of access and egress
- Safe appliances and equipment for doing the work
- A safe system of work
- Competent fellow employees.

Further cases have elaborated on these original basic duties established in 1938. As stated in *Wilson's & Clyde Coal Co. Ltd v English*,

> *The whole course of legal authority consistently recognizes a duty which rests on the employer, and which is personal to the employer, to take reasonable care for the safety of his workmen.*

Since the Employer's Liability (Compulsory Insurance) Act 1969, all employers have a duty to carry insurance against claims for damages made by their employees as a result of injuries and ill health sustained at work.

The Walker decision concerned a local authority, but the same principles outlined above would apply equally to any other employer, for instance, in industrial or commercial workplaces. In this particular case, the potential for a system of work imposed on an employee, such as dealing with child abuse cases, creating stress on that employee, was considered. Was the system of work safe in the light of the previous representations made by Mr Walker to his employers?

The judgement in the Walker case clarified the situation relating to psychiatric injury thus:

> *Whereas the law on the extent of this duty has developed almost exclusively in cases involving physical injury to the employee as distinct from injury to his mental health, there is no logical reason why risk of psychiatric injury should be excluded from the scope of the duty of care.*

The aggressive work culture adopted by many organizations could be a significant factor in establishing breaches of the common duty of care. For example, many employers still see manifestations of stress, such as anxiety or high levels of emotion, as a sign of weakness. They may simply refuse to acknowledge, firstly, the existence of stress and, secondly, that the decisions they make affecting subordinates may induce stress in those people. The duty of care owed to employees does not figure in their deliberations and decisions.

8.2.3 Negligence and stress

Negligence is defined as 'careless conduct injuring another'. The duties of employers at common law (see above) form part of the general law of negligence and, as such, are specific aspects of the duty to take reasonable care.

Negligence has been defined at common law as the existence of a duty of care; breach of that duty; and injury, damage or loss resulting from, or caused by, that breach (*Lochgelly Iron & Coal Co. Ltd v M'Mullan (1934) AC 1*).

Employers should recognize that the duty of care is of a personal nature. In other words, the duty is owed to each employee and not to the workforce as a whole. Thus, in any civil action by an employee against his employer based on the tort of negligence, the court is entitled to examine the circumstances of the individual employee and what it was reasonable to expect of the employer in the particular case.

However, the burden of proof in a claim for negligence falls on the employee, that is, the employee must prove that the employer was at fault or to blame. This is the nature of negligence. As stated in *Lochgelly Iron & Coal Co. Ltd v M'Mullan*,

> *In strict legal analysis, negligence means more than heedless or careless conduct, whether in omission or commission it properly connotes the complex concept of duty of care, breach of that duty of care and the injury suffered as a direct consequence by the person to whom the duty was owing.*

Causation and foreseeability

In actions for negligence, the legal concepts of 'causation' and 'foreseeability' are significant. It is for the claimant to prove that the injury suffered was actually caused by work-related circumstances, and that the employer should reasonably have foreseen that such injury would be caused. What is important is that once the claimant employee has established a breach of duty by the employer, it must be shown that this breach caused, or materially contributed to, the injury suffered. It is not sufficient to show that occupational stress caused the harm. It is the employer's failure to take appropriate action in the light of the fact that psychiatric injury to the claimant was reasonably foreseeable.

Where the psychiatric injury has more than one cause, for example, home-related circumstances, an employer will only be expected to pay for the proportion of the harm suffered which is attributable to his wrongdoing, i.e. his failure to take action. Thus, when assessing damages, a court would take into account any pre-existing condition, disorder or vulnerability and the possibility that the employee would have suffered some form and level of psychiatric injury irrespective of the breach of duty by the employer.

Establishing stress-induced injury (psychiatric injury) requires some form of assessment and analysis by specialists with a view to identifying the causes, both direct and indirect, of the stress. Individual causes of stress in the workplace can be associated with the individual's role in the organization, career development (or lack of it), the organizational structure and climate, factors intrinsic to the job, relationships within the workplace and job or the classic home/work conflict situation (company versus family demands).

Much of this information can be obtained through the use of a well-validated questionnaire, by discussion and consultation with a clinical psychologist, occupational physician or occupational health nurse. In the majority of cases, the individual's general medical practitioner should be involved from the outset and supported by a clinical psychologist.

Failure by an employer to undertake these measures could feature significantly in a claim for negligence by an employee.

Following the precedent set by the Walker case, the number of stress-related legal actions brought before the civil courts has increased year by year.

8.2.4 Breach of statutory duty

In certain cases a breach of a criminal duty imposed by a statute or regulations may give rise to civil liability and a resultant claim for damages. The standard test adopted by the courts has been to ask the question:

> *Was the duty imposed specifically for the protection of a particular class of person, or was it intended to benefit the public at large?*

If the answer to the first part of the question is 'yes', a civil claim may be allowed. The courts have always viewed legislation, such as the safety provisions of the Factories Act 1961, as being directed towards a particular class of person and have allowed civil claims for damages by persons belonging to those protected classes, namely employees.

Under the Management of Health and Safety at Work Regulations (MHSWR), civil liability for a breach of these regulations was initially excluded. However, under the Management of Health and Safety at Work Regulations and Fire Precautions (Workplace) (Amendment) Regulations 2003, this exclusion was modified to apply only in the case of persons *not* in the employment of the employer in question.

The implications of this amendment to the MHSWR are significant in that, in future, employees suffering stress-related ill health may be in a position to sue their employer within the tort of breach of statutory duty, based on the employer's duties under the MHSWR, and with particular significance to the employer's duties under the MHSWR with respect to undertaking risk assessment, implementation of preventive and protective measures arising from the risk assessment process, the installation of health and safety arrangements, health surveillance, information for employees, consideration of human capability and the provision of training (see further Chapter 9). In particular, specific measures may be necessary to protect 'vulnerable' groups, such as new or expectant mothers and young persons, from the risk of stress-related ill health.

Recent Court of Appeal judgement

Recently the Court of Appeal in England issued its judgement in *Chairman of the Governors of St Thomas Beckett RC High School v Hatton* (see later in this chapter) and other related appeals relating to significant awards of damages in stress at work cases. The judgement emphasizes that there is an obligation on employees to complain about stress before it is too late, rather than keeping quiet. On this basis, an employer is entitled to assume that an employee can withstand the normal pressures of the job, unless he knows to the contrary. On the other hand, however, employers are expected to take steps to assess all risks of injury at work (in accordance with the MHSWR), and to act in individual cases where there are obvious indications of excessive stress.

This Court of Appeal decision does shift some of the burden in dealing with stress back on to employees, but all employers remain under a duty of care to take 'reasonable steps' to safeguard employees from psychiatric harm. The judgement suggests that the provision of confidential counselling services and advice, with access to treatment, would be a significant factor in evading liability for stress-related claims.

8.2.5 Reasonably foreseeable injury

Does the doctrine of reasonably foreseeable injury established in *John Summers v Frost (1955) 1 AER 870* apply in the case of stress at work? This case concerned the absolute duty of the employer under section 14 of the Factories Act 1937 to ensure fencing of machinery, in this case, a grindstone, in a factory. Clearly, a parallel can be drawn between exposing employees to risks to their physical safety as opposed to exposing them to risks to their mental health whilst at work.

Before the Walker case, the duty of care did not extend to situations where employees had suffered mental injury, such as that caused by stress at their workplace, as a result of heavy workload commitments imposed upon them by their employers. (*Petch v HM Customs & Excise Commissioners (1993) ICR 789*).

As a result of the Walker case, employers clearly have a responsibility for encouraging an understanding of the causes and effects of stress in the workplace. Whilst it must be recognized that some level of stress exists in workplaces, there is a duty on employers to reduce or minimize these levels of stress. What is important is that, prior to the Walker case, stress at work was perceived as a grey area due to the fact that it is difficult to measure in human terms and that the manifestations and effects of stress vary from person to person.

The importance of the Walker case is that it was the first case in legal history where an employee was awarded damages for psychiatric injury suffered as a result of work-related stress. Previous cases included *Johnstone v Bloomsbury Health Authority*, which was settled out of court, and *Petch v HM Customs & Excise Commissioners*, which was lost by the claimant.

As stated in the judgement in the Walker case:

> *Whereas the law on the extent of this duty has developed almost exclusively in cases involving physical injury to the employee as distinct from injury to his*

mental health, there is no logical reason why risk of psychiatric injury should be excluded from the scope of the duty of care.

The foreseeability test is still the major barrier for an employee. If a claim is to succeed, the claimant must convince the court that it was reasonably foreseeable to the employer that the work would give rise to psychiatric injury in an employee of normal fortitude. Any pre-existing known vulnerability will make this test easier to overcome by the defence. Furthermore, emotions or feelings of frustration, dissatisfaction, upset or embarrassment will be insufficient. It is well recognized that emotional upset or turmoil is a common human trait, and a court will always be reluctant to impose a duty on employers to protect employees from the 'normal stresses' of work.

8.3 Court of Appeal guidelines: Employers' obligations

Court of Appeal guidelines with respect to an employer's obligations, when faced with a reasonably foreseeable risk of illness to an employee, are outlined below:

1. The employer should pay regard to the size of the enterprise and its administrative resources, as there are limits as to what can be reasonably expected of an employee.
2. Before considering what action should be taken, an employer must pay regard to:
   - the magnitude of the risk of psychiatric injury occurring;
   - the seriousness of the potential injury;
   - the cost and practicability of taking action to prevent such an injury; and
   - whether these steps would actually prevent the injury from occurring.
3. Whilst it may be reasonable to consider granting an employee:
   - a sabbatical;
   - transferring him to another department;
   - redistributing work; or
   - granting assistance,

an employer would not be obliged to dismiss or demote an employee in order to remove him from a stressful situation.

Moreover, an employer who provides a confidential support and advice service will be unlikely to be found in breach of duty unless he has placed unreasonable demands upon an individual when the risk of psychiatric injury to that individual was obvious.

8.4 Linking stress to the workplace

As with any aspect of occupational health practice, it is essential that the potential for ill health associated with working conditions, systems, substances and procedures is recognized. Once this potential for ill health has been recognized, it is possible to measure the factors which create the condition and evaluate them against known criteria. This concept applies equally well in the case of physically induced stressors, such as those associated with noise and vibration, as with those of a social or psychologically induced nature, which include the stress-related conditions.

Employers clearly have a responsibility for encouraging an understanding of, and the principal manifestations of stress. Whilst it must be recognized that, inevitably, there will be some level of stress at work, there is a duty on employers to minimize levels of stress as far as possible.

Staff should be encouraged to maximize their potential through appropriate training and development programmes. Many organizations run stress management programmes to assist and advise staff to, firstly, recognize those things in life that cause the individual stress and, secondly, on personal measures they can take to reduce stress, such as progressive muscular relaxation, hypnotherapy, stress counselling and other forms of coping strategy.

8.5 Recent cases

8.5.1 Young v The Post Office (Court of Appeal, 30.4.2002)

The words 'I need help' featured in a memorandum from the claimant to his employers prior to contracting depression. Despite a clear message that he was struggling, the psychiatrists in the case jointly agreed that there was no foreseeable risk of psychiatric illness from the work that he was doing at the time that he wrote the memorandum.

Having had a period off work, and as in *Walker v Northumberland County Council,* being promised a much more pleasant working environment on his return, Mr Young 'cracked' to use the term adopted by Lord Justice May. He succeeded in his claim for damages arising out of the second episode.

This was the first decision from the Court of Appeal in an occupational stress case following the important decision in the cases headed by *Sutherland v Hatton* and the judgement of Lady Justice Hale. The issues in this particular case were:

- Is it the responsibility of the claimant to notify his employer if he is finding the working environment stressful or cannot cope?
- Can the claimant be guilty of contributory negligence for failing to bring his circumstances to the attention of his employer?

With regard to the first question, the defendants had offered what seemed like a remarkably flexible environment to which he could return. He could work the hours that he wished and go out for a walk if he chose. In fact, he was not obliged to do any work at all!

However, the employers did not monitor what happened to Mr Young on his return and did not assess his suitability and prevent him from going on a training course, which he found extremely stressful. Additionally, he ended up, through the absence of another manager, back in the management role that had caused his problems initially.

The court did not criticize Mr Young for failing to mention his increasing stress levels to his employer because he was accepted as a conscientious worker who would endeavour to do the best he could. The court held that he should not be criticized for doing so. Lord Justice May found, that while accepting Lady Justice Hale's view that 'because of the very nature of psychiatric disorder, it is bound to be harder to foresee than physical injury', it was 'plainly foreseeable that there might be a recurrence if appropriate steps were not taken when he returned to work. The employers owed him a duty to take such steps.'

With respect to the allegation of contributory negligence, the defendants contended that Mr Young created his own burden by undertaking stressful work and choosing not to notify his employers that he found it so. In dismissing the allegation, Lord Justice May said that while it was 'theoretically possible' for a finding of contributory negligence in cases of psychiatric illness, it would be unusual. Employers cannot expect an employee who is known to be vulnerable to be responsible for recurrent psychiatric illness, even if he fails to tell them that his job is once more becoming stressful.

The decision offers some hope for cases in which the claimant suffers a second bout of illness (a *Walker* claimant). It demonstrates that despite *Sutherland,* there are instances in which the burden has not completely shifted on to the mentally ill claimant to resolve his difficulties.

The lesson to be learnt from this case is that an employer, knowing of an employee's mental state, is under an obligation to not only design a less stressful working environment, but to ensure that it is implemented.

8.5.2 Lancaster v Birmingham City Council

A former employee of Birmingham City Council was awarded £67 000 in damages for stress that forced her to retire as a council housing officer at the age of 41. Beverley Lancaster worked for the council for 26 years, mainly in clerical and technical roles. When her post was abolished she was appointed as a housing officer, a post for which she had neither experience nor qualifications.

She soon found that she was not being given the support and training she had been promised and could not cope with the demands of her new job. As a result, during the subsequent 4 years, she was off work sick for half that time. During the first period of illness, which lasted 3 months, she was diagnosed as suffering from depression. Later, a consultant diagnosed her as suffering from work-related stress and she took longer periods of sick leave.

She stated that her employer did not take her concerns seriously. 'I was never treated as an individual. They never took my illness seriously, although I had approached them on many occasions for medical support', she said.

In this case, in which the claimant was supported by the public sector trade union, UNISON, the employer actually admitted liability for the stress caused at work.

8.5.3 Terence Sutherland (Chairman of the Governors of St Thomas Beckett RC High School) v Hatton and other related appeals

Penelope Hatton taught French at a comprehensive school at Huyton, Merseyside, for 15 years until she left because of depression and debility. She was awarded £90 765 damages after suing St Thomas Beckett RC High School.

In February 2002, the Court of Appeal issued its judgement in respect of the above appeal and other related appeals relating to significant awards of compensation in stress at work cases. Allowing the school's appeal, the judges said her workload was no greater than other teachers in a similar school. Her pattern of absence and illness were 'readily attributable' to causes other than work stress.

Generally, an employer is entitled to assume that the employee can withstand the normal pressures of the job, unless he knows to the contrary. On the other hand, employers are expected to take steps to assess all risks of injury at work and to act in individual cases where there are obvious indications of excessive stress.

The Court of Appeal's decision does shift some of the burden in dealing with stress at work back on to the employee, but all employers remain under a duty of care to take 'reasonable steps' to safeguard employees from harm.

The judgement suggests that the provision of a statement of policy on stress at work and formal assessment of stress-related risk, supported by confidential counselling services from, for instance, an occupational health service, with ready access to treatment, would be a significant factor in evading some liability for stress-related ill health.

8.5.4 Somerset County Council v Leon Alan Barber

Alan Barber was Head of Mathematics for East Bridgewater Community School for over 12 years. He sued Somerset County Council and was awarded £101 041 after developing symptoms of depression, brought on by regularly working up to 70 hours a week, leading to his losing control of the classroom.

The Court of Appeal allowed the council's appeal on the basis that Mr Barber had not told senior management at the school about his symptoms until his breakdown. However, the House of Lords ruled that employers must generally take the initiative to protect employees, particularly in the case where they know an individual is vulnerable to stress-related illness. Mr Barber was subsequently awarded £72 547 plus interest and costs.

This ruling fundamentally overturned the Court of Appeal verdict that his employers, Somerset County Council, had not breached their duty of care towards him and were not required to ensure that he was not still suffering stress-related ill health after the summer school holiday, despite being aware of his condition the previous term.

In his judgement, which was backed by three of the four other Law Lords, Lord Walker said:

> *At the very least the senior management team should have taken the initiative in making sympathetic enquiries about Mr Barber when he returned to work, and making some reduction in his workload to ease his return.*
>
> *Even a small reduction in his duties, coupled with the feeling that the senior management team was on his side, might by itself have made a real difference. In any event, Mr Barber's condition should have been monitored, and if it did not improve, some drastic action would have had to be taken.*

Whilst broadly supporting the Court of Appeal's verdict that employers are only liable if they know an employee is at risk, Lord Walker said that employees need not be forceful in their complaints, or go into detail, particularly as depression makes it more difficult to complain.

This decision has implications for employers. Firstly, they must take the initiative where employees report stress arising from work. Secondly, they must keep up-to-date with the current approaches to stress and take measures to reduce the risk of injury from same, as they do in the case of risk of physical injury. Thirdly, it is conceivable that more civil claims of this nature are likely to be successful where employers have failed to follow best practice and employees have suffered psychological injury as a result.

On the other hand, the decision in this case does not lay down in tablets of stone what both employers and employees need to do in cases of work-related stress. Furthermore, it does not alter the guidelines set out by the Court of Appeal in *Sutherland v Hatton* in terms of when the duty of care arises which is, of course, a matter of foreseeability.

What this decision does do, however, is to provide a broad view as to what an employer should do once a duty to take reasonable care has become evident. In particular, an employer cannot assume that, because an employee, who formerly claimed to be suffering from stress, returns to work and makes no further complaint, that he has discharged his duty towards that employee.

Clearly, under the civil law, the duty of care owed by an employer to an employee does not arise unless the manifestations of future harm to health are clear enough for a reasonable employer to recognize that he needs to take some form of action. The claimant, Mr Barber, had, by July 1996, done enough to satisfy this test. His employer should then have taken some form of positive action to remedy the situation.

8.5.5 Baker Refractories Ltd v Melvyn Edward Bishop

Melvyn Bishop worked in the factory until 1997 when he had a breakdown and attempted suicide. He was dismissed in 1998 after failing to return to work. He subsequently sued the company and was awarded £7000 damages in 2001.

The judges, allowing the appeal, said there was nothing excessive about the demands of his work in this case.

8.5.6 Sandwell Metropolitan Borough Council v Olwen Jones

Ms Jones was an administrative assistant until 1995 and sued Sandwell MBC when she fell ill with anxiety and depression. She was awarded £157 541 after a County Court judge was told she had to work 'grossly excessive hours' and do the work of three people.

The judges, dismissing the appeal, said they had to decide whether it was foreseeable by the employers that such unreasonable demands would cause harm.

8.5.7 Benson v Wirral Metropolitan Borough Council

In this case no liability was admitted. However, the teacher concerned was awarded £47 000 in an out-of-court settlement.

8.5.8 Ingram v Hereford and Worcester County Council

This case was settled out of court, the County Council admitting liability. Here a warden employed by the council to oversee a travellers' site frequently had his decisions undermined by other council officers. This damaged his working relationships with the residents, resulting in both physical and verbal abuse. He eventually gave up work because of ill health.

He was awarded £203 000, a record amount for work-related stress.

8.5.9 McLeod v Test Valley Borough Council [2000, unreported]

Mr McLeod took an action against his employers, Test Valley Borough Council, for chronic depression and persistent delusional disorder arising from workplace bullying and harassment from his manager.

This case was settled out of court, with no liability admitted, in the sum of £200 000.

8.5.10 Rorriston v West Lothian College [1999 LR 102]

Mrs Rorriston sought damages following severe anxiety and depression caused through the constant harassment and humiliation she suffered from her line manager and another colleague. The Outer House of the Court of Session dismissed this claim, even though Mrs Rorriston was suffering severe psychological stress, as she was not suffering from 'a recognized psychological disorder'.

In his judgement, Lord Reed said:

> *Many if not all employees are liable to suffer those emotions (i.e. to be unsatisfied, frustrated, embarrassed and upset) and others mentioned in the case, such as stress, anxiety, loss of confidence and low mood. To suffer such emotions from time to time, not least because of problems at work, is a normal part of human experience. It is only if they are liable to be suffered to such a pathological degree as to constitute a psychiatric disorder that a duty of care to protect against them can arise; and that is not a reasonably foreseeable occurrence unless there was some specific reason to foresee it in a particular case.*

The outcome of this ruling is that a claimant must now demonstrate a higher standard of proof of an employer's breach of his duty of care. Generally, the emotional consequences of exposure to stress are insufficient. Some recognized and recognizable psychiatric illness must have resulted or be about to result.

8.5.11 Maryniak v Thomas Cook Ltd [Unreported, 20.10.1998]

Maryniak was employed as branch manager at Thomas Cook's offices. During his period of employment work systems changed, which the claimant was reluctant to accept, and showing a high level of inflexibility in taking on the new regime. This performance was eventually reflected in two successive performance appraisals and he was demoted. Mr Maryniak subsequently took a period of sick leave for depression and was eventually dismissed.

His contention was that his employer had known of his depressive state owing to earlier visits to his doctor. However, Thomas Cook indicated, that whilst they were aware of the fact that he was receiving treatment, they were unaware of the actual reason for it. There was some discrepancy between the doctor's initial diagnosis, namely of a non-work related viral infection, and he did not, after the next consultation, record the diagnosis as one of depression.

The sickness absence continued. However, Maryniak's conduct throughout 1993 could not have suggested to his employer that he was suffering from work-related psychiatric illness and, as a result of his employer's lack of awareness of the true cause of sickness absence, there was held to be no negligence. Maryniak's claim for damages failed.

This case reinforced the need for an employee to ensure his employer is made aware of any psychiatric disorder arising from work. An employer cannot 'reasonably foresee' that stressors arising from work could have an injurious effect on employees that may result in a recognized psychiatric condition.

8.6 Practical propositions

In 2002, after hearing four appeal cases, namely:

1. *Terence Sutherland (Chairman of the Governors of St Thomas Beckett RC High School) v Penelope Hatton*
2. *Somerset County Council v Leon Alan Barber*
3. *Sandwell Metropolitan Borough Council v Olwen Jones*
4. *Baker Refractories Ltd v Melvyn Edward Bishop*

covered earlier in this chapter, Lady Justice Hale, sitting with Lords Justice Brooke and Kay, outlined a number of 'practical propositions' to assist courts in dealing with future claims for psychiatric injury arising from stress at work.

1. There are no special control mechanisms applying to claims for psychiatric (or physical) illness or injury arising from the stress of doing the work the employee is required to do. The ordinary principles of employer's liability apply.
2. The threshold question is whether this kind of harm to an employee was reasonably foreseeable. This has two components:
   - an injury to health (as distinct from occupational stress); and
   - which is attributable to stress at work (as distinct from other factors).
3. Foreseeability depends upon what the employer knows (or ought reasonably to know) about an employee. Because of the nature of a mental disorder, it is harder

to foresee than physical injury, but may be easier to foresee in a known individual than in the population at large. An employer is usually entitled to assume that the employee can withstand the normal pressures of the job unless they know of some particular problem or vulnerability.

4. The test is the same whatever the employment. There are no occupations that should be regarded as intrinsically dangerous to mental health.
5. Factors likely to be relevant in answering the threshold question include:
   - **The nature and extent of the work done by the employee**: Is the workload much more than is normal for the particular job? Is the work particularly intellectually or emotionally demanding for this employee? Are demands being made of this employee unreasonable when compared with the demands made of others in the same or comparable jobs? Or are there signs that others doing this job are suffering harmful levels of stress? Is there an abnormal level of sickness or absenteeism in the same job or in the same department?
   - **Signs from the employee of impending harm to health**: Have they a particular problem or vulnerability? Have they already suffered from illness attributable to stress at work? Have there recently been frequent or prolonged absences which are uncharacteristic? Is there reason to think that these are attributable to stress at work, for example because of complaints or warnings from them or others?
6. The employer is generally entitled to take what they are told by their employee at face value, unless they have good reason to think to the contrary. They do not generally have to make searching enquiries of the employee or seek permission to make further enquiries of their medical advisers.
7. To trigger a duty to take steps, the indications of impending harm to health arising from stress at work must be plain enough for any reasonable employer to realize that they should do something about it.
8. The employer is only in breach of duty if they have failed to take the steps which are reasonable in the circumstances, bearing in mind the magnitude of the risk of harm occurring, the gravity of the harm which may occur, the costs and practicability of preventing it, and the justifications for running the risk.
9. The size and scope of the employer's operation, its resources and the demands it faces are relevant in deciding what is reasonable; these include the interests of other employees and the need to treat them fairly, for example, in any redistribution of duties.
10. An employer can only reasonably be expected to take steps that are likely to do some good. The court is likely to need expert evidence on this.
11. An employer who offers a confidential advice service, with referral to appropriate counselling or treatment services, is unlikely to be found in breach of duty.
12. If the only reasonable and effective step would have been to dismiss or demote the employee, the employer will not be in breach of its duty in allowing a willing employee to continue in the job.
13. In all cases it is necessary to identify the steps that the employer both could and should have taken before finding them in breach of their duty of care.
14. The claimant must show that breach of duty has caused or materially contributed to the harm suffered. It is not enough to show that occupational stress has caused the harm.

15. Where the harm suffered has more than one cause, the employer should only pay for that proportion of the harm suffered which is attributable to their wrongdoing, unless the harm is truly indivisible. It is for the defendant to raise the question of apportionment.
16. The assessment of damages will take account of any pre-existing disorder or vulnerability and of the chance that the claimant would have succumbed to a stress-related disorder in any event.

Ex-gratia payments

In 1996, following the riot in Strangeways prison, seven prisoners were awarded nearly £5000 by the Home Office after claiming that their personalities changed because of the riot. The prisoners alleged they suffered post-traumatic stress disorder after witnessing the riot in 1990. Some of the claimants were in segregation under Rule 43, used to protect sex offenders, informers and other inmates thought to be at risk.

One prisoner claimed that the Home Office was in breach of its duty of care to him as a prisoner because it failed to prevent the riot. In this case, he tried, but failed, to rescue two other prisoners from a burning cell. This experience, he alleged, changed him from 'a happy-go-lucky person' to 'a time bomb about to go off'. He said his relationship with the mother of his two children broke down and he suffered nightmares and flashbacks. Later, he was one of the prisoners to be brought out of the prison and sedated at Risley Remand Centre, but never received counselling.

8.7 Violence, harassment and bullying at work

Employers need to know what is going on in their workplaces. They cannot expect line managers to keep them informed of adverse situations or activities on an on-going basis. The concern about bullying in the workplace has increased rapidly in recent years. Bullying can take many forms, ranging from physical acts of aggression to aggressive management style and in the unfair or inappropriate allocation of tasks.

Whilst there is no specific criminal legislation dealing with these matters, the employment relationship, that is, the relationship between an employer and an employee, is governed by both the law of contract and the law of tort. In the former an action may lie in breach of contract, in the latter, an action based on negligence. Thus, a contract of employment contains a range of express and implied terms, the latter implied by both common law and statute. Of particular importance in a contract of employment is the concept of 'the duty of care'. Breaches of this duty of care may lead to both a criminal prosecution and/or a civil action in the civil courts and tribunals.

However, bearing in mind the recent developments in current civil case law on stress at work, an employee specifically complaining to a civil court of psychological violence, associated with bullying or harassment on which the employer had singularly failed to take action to prevent it, would have no difficulty in proving the employer's liability for psychiatric illness (see below) and reference would inevitably be made to the Court of Appeal guidelines shown below.

8.8 Court of Appeal general guidelines

The Court of Appeal judges issued the following guidelines:

- The ordinary principles of employer's liability apply to claims for psychiatric illness arising from stress at work.
- Employers are generally entitled to accept at face value what employees tell them, unless there is good reason not to do so.
- No occupation should be regarded as intrinsically dangerous to mental health.
- An employer is entitled to assume that an employee is able to withstand the normal pressures of the job unless they are aware of some particular problem or vulnerability.
- For an employer to have a duty to take action, indications of impending harm to health must be apparent enough to show that the action should be taken.
- An employer is only in breach of duty if he has failed to take steps that are reasonable in all the circumstances.
- An employer who makes available a confidential counselling service with access to treatment is unlikely to be found in breach of duty.

8.8.1 Liability for psychiatric illness

In order to prove an employer's liability for psychiatric injury, a claimant employee is required to demonstrate the following:

- That the employer has breached the duty of care owed to the employee to provide a safe place of work and to keep him safe from harm
- That the employee is suffering from recognizable psychiatric illness, that is, not simply stress but clinical depression or post-traumatic stress disorder
- That the recognizable psychiatric injury was caused by the employer's negligence and not by any other factors
- That the stress to which the employee was exposed was sufficient to create 'a reasonably foreseeable risk of injury'.

8.8.2 The two main issues

In the leading judgement, the Court of Appeal held that liability depends on two main issues:

1. Was it reasonably foreseeable that the employee would suffer psychiatric illness as a result of the working conditions?
2. If so, did the employer do all he could reasonably do to address the problem?

These two questions are significant.

8.9 Establishing stress-induced injury

Establishing stress-induced injury in a civil court is difficult. It is fundamental to a personal injury claim that there is 'injury'. Injury does not necessarily have to be physical injury, but a claim of 'stress' is insufficient. Fundamentally, the courts look for evidence of a clinically recognizable psychological or psychiatric condition.

A number of other factors are also important, namely:

8.9.1 Duty of care

In the majority of stress-related claims by an employee against his or her employer there automatically exists a duty of care. However, there must be clear-cut evidence of a breach of that duty.

8.9.2 Negligence

Establishing negligence on the part of an employer, namely fault or careless conduct, is a question of fact. The employer should have been able to reasonably foresee a risk of injury arising from the work carried out by the claimant.

8.9.3 Medical causation

Proving that the injury was specifically caused by work or conditions of work, as opposed to personal circumstances or personal characteristics, is the major hurdle and where claims for work-related stress can fail.

8.10 Liability for psychiatric illness

The Court of Appeal has indicated a number of factors which need to be considered by employers.

8.10.1 The actual work

- The nature and extent of the work done by the employee.
- Whether the employee's workload is much greater than normal for the kind of job he performs.
- Whether the employee's work is particularly intellectual or demanding.
- Whether the demands made on the employee are unreasonable compared with others in comparable jobs.

- Whether there are signs that others doing the same work are suffering harmful levels of stress.
- Whether there is an abnormal level of absenteeism or sickness in the employee's department.

8.10.2 Areas for investigation by the employer and possible compensation

A second group of factors, in the court's opinion, should further be considered, namely those concerning what the employer knew, or ought reasonably to have known, about the circumstances of the individual employee in question. The court stated that the following issues may be relevant:

- whether there are signs from the employee of impending harm to health;
- whether the employee had a particular problem or vulnerability;
- whether the employee had previously suffered from any illness arising from stress at work;
- whether the employee had taken frequent or prolonged absences from work which were uncharacteristic and which may reasonably cause the employee to consider that they were related to stress at work.

8.11 The remedies for employers

As a result of this High Court decision, employers need to consider a number of questions:

1. Do we recognize the existence of stress?
2. Do we recognize that the decisions and actions that we take may result in stress and stress-related ill health amongst our employees?
3. Is there evidence to suggest the presence of stress in certain employees who may have higher than average sickness absence rates?
4. Is it possible to reorganize work practices and systems to eliminate or reduce the stress on certain employees who may be displaying manifestations of stress at work?
5. Is there a company policy on stress at work, supported by information, instruction and training of staff, together with other forms of support, such as counselling on stress-related issues?
6. Would an occupational health service be of assistance in dealing with this problem?

8.12 A corporate strategy

A corporate strategy for dealing with the problem of stress at work takes a number of stages as follows.

8.12.1 Recognition/Identification of stressors

The problem of stress at work may be identified through a number of sources including:

1. Staff complaints
2. Job descriptions
3. Work-related ill health and sickness absence statistics
4. Performance appraisals/job and career reviews
5. Reports from occupational health practitioners
6. Overtime reports and
7. Accident investigations.

8.12.2 Measurement of stress

Measurement of the extent of the stress may be achieved through techniques such as counselling, the use of health questionnaires and interview by occupational health practitioners.

8.12.3 Evaluation of the stress

The Holmes–Rahe Scale of Life Change Events is frequently used as a starting point in the evaluation of stress.

8.12.4 Eliminating or reducing and controlling stress

In some cases it may be possible to eliminate the stressor completely through change in work practice, relocation of the individual or some form of therapy. In other cases, the stress could be reduced and controlled through job rotation, redesign of the working practices, modified behaviour on the part of an individual or individuals, a reduction in working hours, a change in responsibility and/or increasing participation in the decision-making process. Much will depend on the identified causes of stress.

8.12.5 Monitoring and review

It is essential that the individual be monitored by an occupational health practitioner on an on-going basis to ensure that his ability to cope has improved as a result of the measures taken and that there has been no deterioration in terms of a reversion to stress-related behaviour and ill health. This may take the form of monthly consultations, supported by a number of stress-relieving techniques and therapies.

8.13 Conclusion

Following the precedent set in *Walker v Northumberland County Council*, the number of stress-related actions brought before the civil courts has increased dramatically year by year. According to the Trades Union Congress report *Focus on Union Legal Services* (1998), for example, there were 459 cases of work-related stress in progress through the courts in 1997. This figure rose to 783 cases in 1998.

Employers simply cannot afford to disregard the potential for stress amongst employees if they are to avoid future civil actions.

Questions to ask yourself after reading this chapter

Employers should undertake and maintain the following, bearing in mind the publication of recent Court of Appeal guidelines on the subject:

- Do you appraise employees' workloads on a regular basis?
- Do you take all complaints of overwork and stress seriously, regardless of how trivial they may seem?
- Do you appraise sickness absence records on a regular basis? (Unusual levels of sickness absence may be an indicator of stress in the workplace.)
- Do you conduct return-to-work discussions with employees following extended periods of sick leave?
- Do you seek an opportunity to discuss their concerns or problems?
- Have you established a confidential advisory service, including counselling or treatment services, for vulnerable employees?
- Do you incorporate discussions on stress in the health and safety consultation process with employees?
- Do you keep records of all discussions with individuals on the subject?

Key points – implications for employers

- At common law employers have a duty to take reasonable care to protect the health of employees, including their mental health, from risk of foreseeable injury.
- A failure to protect the mental health of an employee by an employer could result in a civil claim for negligence by the affected employee.
- There is an obligation on employees to complain formally of stress at work to their employer.
- The Court of Appeal has laid down guidelines for employers with respect to their obligations when faced with a reasonably foreseeable risk of mental illness to employees.

- Employers have a responsibility for encouraging an understanding of stress at work and the principal manifestations of stress amongst employees.
- Recent cases in the civil courts have reinforced the need for well-developed management systems for preventing and dealing with stress at work.
- The High Court has produced a number of 'practical propositions' to assist courts in dealing with future stress-related claims.
- An employer who provides a confidential counselling service is unlikely to be found in breach of the duty of care.

9

The criminal implications

The duties of employers to protect the health of their employees are contained in the Health and Safety at Work etc. Act 1974 and subordinate legislation, such as the Management of Health and Safety at Work Regulations 1999 and the Health and Safety (Display Screen Equipment) Regulations 1992.

The principal features of this legislation are outlined below:

9.1 Health and Safety at Work etc. Act 1974 (HSWA)

Section 2: Duties of employers

1. It shall be the duty of every employer to ensure, so far as is reasonably practicable, the health, safety and welfare at work of all his employees.
2. Without prejudice to the generality of an employer's duty under the preceding sub-section, the matters to which that duty extends include in particular:
   - the provision and maintenance of plant and systems of work that are, so far as is reasonably practicable, safe and without risks to health;
   - arrangements for ensuring, so far as is reasonably practicable, safety and absence of risks to health in connection with the use, handling, storage and transport of articles and substances;
   - the provision of such information, instruction, training and supervision as is necessary to ensure, so far as is reasonably practicable, the health and safety at work of his employees;
   - so far as is reasonably practicable as regards any place of work under the employer's control, the maintenance of it in a condition that is safe and without risks to health and the provision and maintenance of means of access to and egress from it that are safe and without such risks;

- the provision and maintenance of a safe working environment for his employees that is, so far as is reasonably practicable, safe, without risks to health, and adequate as regards facilities and arrangements for their welfare at work.

9.2 What is health?

Health is not defined in the HSWA or in any regulations made under the HSWA. One definition of health is 'a state of physical and mental wellbeing' (International Labour Organization). Under the HSWA and subordinate legislation the criminal liability of an employer in terms of protecting the health of an employee is clear, both generally and specifically.

9.2.1 Implications for employers

The duties of employers towards their employees under the HSWA are well established, and employers can be prosecuted where there is evidence of a breach of section 2 which has resulted in the ill health of an employee. That ill health could be caused by failure to adopt an appropriate safe system of work which takes into account potential psychological stressors or the failure to provide a healthy working environment which could result in some form of stress-related ill health.

Similarly, employers may be prosecuted where, in the opinion of an enforcement officer, there may be risk of employees or others contracting an occupational disease due to exposure to physical phenomena, such as noise, chemical agents, such as solvents; or biological agents, such as harmful bacteria. Clearly, the burden of proof rests with the enforcing authority to prove beyond reasonable doubt that a risk to health existed at the time or still exists.

These duties under the HSWA have been greatly strengthened as a result of certain health-related legislation, such as the Noise at Work Regulations and the Control of Substances Hazardous to Health (COSHH) Regulations, where the need to assess the risk to health is a general requirement under these regulations. Generally, the increased public awareness of specific conditions, such as work-related upper limb disorders and legionnaire's disease have, in the last decade, required the enforcement authorities to pay greater attention to health risks in the workplace.

9.3 Management of Health and Safety at Work Regulations 1999 (MHSWR)

These regulations, in the main, place absolute or strict duties on employers, compared with HSWA, where the duties are qualified by the term 'so far as is reasonably practicable'.

Regulation 3: Risk assessment

1. Every employer shall make a suitable and sufficient assessment of
   - the risks to the health and safety of his employees to which they are exposed whilst at work; and
   - the risks to the health and safety of persons not in his employment arising out of or in connection with the conduct by him of his undertaking, for the purpose of identifying the measures he needs to take to comply with the requirements and prohibitions imposed upon him by or under the relevant statutory provisions and by Part II of the Fire Precautions (Workplace) Regulations 1997.

Regulation 4: Principles of prevention to be applied

Where an employer implements any preventive and protective measures he shall do so on the basis of the principles specified in Schedule 1 of these Regulations (see below).

Schedule 1: General principles of prevention

(*This Schedule specifies the general principles of prevention set out in Article 6(2) of Council Directive 89/391/EEC*)

- avoiding risks;
- evaluating the risks which cannot be avoided;
- combating the risks at source;
- adapting the work to the individual, especially as regards the design of workplaces, the choice of work equipment and the choice of working and production methods, with a view, in particular, to alleviating monotonous work and work at a predetermined work-rate and to reducing their effect on health;
- adapting to technical progress;
- replacing the dangerous by the non-dangerous or the less dangerous;
- developing a coherent overall prevention policy which covers technology, organization of work, working conditions, working relationships and the influence of factors relating to the working environment;
- giving collective protective measures priority over individual protective measures; and
- giving appropriate instructions to employees.

Regulation 5: Health and safety arrangements

1. Every employer shall make and give effect to such arrangements as are appropriate, having regard to the nature of his activities and the size of his undertaking, for the effective planning, organization, control, monitoring and review of the preventive and protective measures.
2. Where the employer employs five or more people, he shall record the arrangements referred to in paragraph (1).

Regulation 6: Health surveillance

Every employer shall ensure that his employees are provided with such health surveillance as is appropriate having regard to the risks to their health and safety which are identified by the assessment.

Regulation 10: Information for employees

Every employer shall provide his employees with comprehensible and relevant information on:

- the risks to their health and safety identified by the assessment;
- the preventive and protective measures, etc.

Regulation 13: Capabilities and training

1. Every employer shall, in entrusting tasks to his employees, take into account their capabilities as regards health and safety.
2. Every employer shall ensure that his employees are provided with adequate health and safety training:
   - on their being recruited into the employer's undertaking;
   - on their being exposed to new or increased risks because of
     (a) their being transferred or given a change of responsibilities within the employer's undertaking;
     (b) the introduction of new work equipment into or a change respecting work equipment already in use within the employer's undertaking;
     (c) the introduction of new technology into the employer's undertaking; or
     (d) the introduction of a new system of work into or a change respecting a system of work already in use within the employer's undertaking.

Note: These regulations brought in a totally new approach to dealing with the risks arising from work compared with the requirements of the HSWA. Any consideration of the duties listed above would indicate the need for employers to consider the potential for stress-related ill health arising from their work activities. Reference must be made to the 'General Principles of Prevention' outlined in Schedule 1, in particular, sub-paragraph (d).

9.4 Management of Health and Safety at Work and Fire Precautions (Workplace) (Amendment) Regulations 2003

These regulations modified the above regulations by restricting civil liability for a breach of statutory duty.

In the case of the Management of Health and Safety at Work Regulations, regulation 22 was amended whereby 'breach of a duty imposed on an employer by these Regulations shall not confer a right of action in any civil proceedings insofar as that duty applies for the protection of persons not in his employment'. This restriction to persons *not* in the employment of an employer, implies that employees may be in a position in future to instigate civil proceedings in respect of a breach of duty laid down in the Regulations, particularly in the case of those Regulations outlined above.

9.5 HSE stress questionnaire

HSE guidance identifies seven stressors which are shown as a form of questionnaire, thus:

- **Demands**: Are staff comfortable with the amount of work they have to do or the hours they are expected to work?
- **Control**: Are staff involved in deciding what work they do, and when and how they do it?
- **Support**: Are you offering adequate managerial support to staff, for example, with new work, with everyday issues or if they experience personal problems? Are all staff properly trained in the tasks they are expected to perform?
- **Relationships**: How are relationships conducted in your workplace? Are there problems with bullying or harassment?
- **Roles**: Are staff clear about what is expected of them? Are staff struggling with multi and/or conflicting roles?
- **Change**: Do you communicate and consult adequately with your staff about organizational change?
- **Culture**: Do you promote open dialogue between staff and managers? Do staff feel the organizational culture is sexist or discriminatory towards ethnic minorities?

9.6 Stress and risk assessment

Stress is a risk to which many people are exposed, particularly those dealing with the public, such as teachers, nurses and doctors. However, others in less exposed work situations may experience stress, the causes of which may not be obvious.

Analysis of sickness absence records could well be the starting point for assessing whether certain employees are more exposed to stress than others, in particular, those employees with identified frequent periods of short-term sickness absence. Other factors which are significant are high levels of employee turnover in certain parts of an organization and evidence of disciplinary problems.

In order to identify, measure and evaluate the risk to the health of employees, staff can be asked to complete a Stress Audit or Questionnaire (see page 213) which, preferably, is tailored around the organization's activities. One of the problems, however, with this measurement technique is that many people see the admission of stress as a sign of weakness which, they fear, could be used against them by management. On this basis, anyone being asked to complete the audit should be reassured that the information is confidential. In certain cases, it would be appropriate for this information to be obtained on an anonymous basis.

As with any form of risk assessment, the recommendations for preventing or controlling exposure should outline actions to be taken on a short-, medium- and long-term basis. These actions may include modifications to management practices and human resources policies, the redesign of work practices to include more flexible shift working, the provision of increased information and training and, in some cases, on-going support through regular counselling and health surveillance measures.

Once these measures have been implemented, some form of monitoring and evaluation is necessary to ensure continued support from management and the implementation of the changes recommended. Records should be maintained of action taken.

9.6.1 HSE guidance on stress risk assessment

The HSE has published the following guidance on stress risk assessments.

Step 1: Identify the hazards

Good sickness absence data monitoring is the key. If a particular team or unit has high sickness absence, investigate the causes. Conditions or work organization may be raising stress levels and, in turn, sickness.

Conduct return-to-work interviews to find out why staff are taking time off for stress.

Talk to your staff and get them to talk to you. You don't need to mention the word 'stress'. It can be easier just to ask staff about things that upset them or make work difficult.

Use focus groups to get staff to talk about stress and bounce solutions off each other.

Conduct exit interviews if staff turnover is high.

Step 2: Establish who might be harmed and why

Use the seven stressors to group the issues identified in step 1 roughly under headings. This is a useful first step in sorting and prioritizing the information you have obtained.

Step 3: Develop an action plan

You can't tackle everything your risk assessment identifies at once. Smaller problems which can be solved quickly (for example, improving communication by introducing regular team meetings) are good things to start with. This should immediately reduce overall stress levels, making it easier to solve more difficult problems over time. It also demonstrates that you are serious about tackling the problems.

When contemplating more costly measures (i.e. employing extra staff), consider whether the potential benefits justify the financial cost.

Consult and involve staff when deciding what to do.

Step 4: Take action

You must make practical interventions to reduce the exposure of your employees to the stressors identified as presenting the greatest risk.

There is no 'one size fits all' solution to each stressor. Take a look at HSE's guidance. How much are you doing towards the guidelines set out there?

Look at what other organizations, in particular similar organizations, have done or are doing.

Step 5: Evaluate and share your work

Try to demonstrate quantitative improvements, for example, a reduction in staff turnover.

After each action, repeat step 1 to establish whether staff feel any of the problems have been reduced or eliminated.

9.6.2 Recording the assessment

In accordance with the Approved Code of Practice to the Management of Health and Safety at Work Regulations, an employer with five or more employees must record the 'significant findings' of the risk assessment. The significant findings of the risk assessment should include:

- a record of the preventive and protective measures in place to control the risks;
- what further action, if any, needs to be taken to reduce risk sufficiently; and
- proof that a suitable and sufficient risk assessment has been made.

In many cases, employers will also need to record sufficient detail of the assessment itself, so they can demonstrate (e.g. to an inspector or to safety representatives or other employee representatives) that they have carried out a suitable and sufficient assessment. This record of the significant findings will also form a basis for revision of the assessment.

9.6.3 The significant findings of a stress risk assessment

The written record of the significant findings of a stress risk assessment must be produced in accordance with the three criteria listed above. Typical examples of the information recorded would include:

- **Preventive and protective measures**: The installation of a stress management programme, a written policy on stress at work, provision of counselling in appropriate cases, on-going consultation with employees, the provision of information, instruction and training for employees and health surveillance by an occupational health nurse are all examples of preventive and protective measures.
- **Further action**: Examples include steps to prevent or reduce bullying or harassment of employees, measures to protect employees who may have to deal with members of the public and holding regular staff meetings to discuss stress-related issues.
- **Proof of a suitable and sufficient risk assessment**: This may be provided by a record of the outcome of a survey amongst employees using a well-designed stress questionnaire, the analysis of the comments made by employees in the questionnaire and measures to be considered by management.

9.6.4 Questions

A number of questions need to be considered when undertaking a stress risk assessment and in considering the significant findings of the assessment.

1. Does the risk assessment adequately consider the potential for stress and stress-related ill health?
2. Do the management systems introduced consider this matter?
3. What provision is made for the health surveillance of staff identified from the risk assessment who may be subject to stress in their day-to-day activities, such as those dealing with members of the public, or in the case of social workers, dealing with child abuse cases?
4. What is the quality and extent of information given to employees on the potentially stressful aspects of their jobs?
5. From a human capability viewpoint, who decides that a particular individual has 'got what it takes' to cope with the stresses associated with various jobs? What are the factors that an employer would consider at the selection stage and subsequently where an employee may be complaining of stress and seeking alternative work?
6. Bearing in mind the duty to train employees under the MHSWR, what measures are taken from a training viewpoint to alert people to the potentially stressful situations that can arise at work, such as dealing with members of the public?
7. Does the organization provide confidential advice to employees, together with counselling and other assistance, on stress-related issues?

The above questions raise a whole series of issues for employers in terms of ensuring employees are not exposed to stress which could result in any of the stress-related conditions and the ill health associated with such conditions. These issues should be recorded in the risk assessment as part of the significant findings of the risk assessment.

One final question needs to be asked, namely:

Is the employer, or member of the management team, who is charged with the task of undertaking stress risk assessments, sufficiently competent to undertake this task?

This sort of assessment requires a reasonable understanding of human psychology, the causes and effects of stress and forms of psychosocial hazard. Simply undertaking a stress questionnaire exercise may not deliver the accurate information that organizations require if they are to assess whether some form of intervention is required and what is the best approach to suit the particular problems identified. Inevitably, this question identifies a need for training of the risk assessor prior to undertaking this exercise.

An organization, on the other hand, may seek the guidance of consultants who specialize in this form of work. It is imperative that such consultants are competent and adequately qualified.

9.6.5 Enforcement

It is conceivable that, as a result of stress-related ill health being well established in the civil courts, enforcement officers are now in a position to take action under the

HSWA. This could be through the service of an improvement notice under section 21 of the HSWA on an employer to, for instance, install a stress management programme or take other actions, such as risk assessment, with respect to stress at work.

Where an inspector considers certain activities will involve 'a risk of serious personal injury', which includes psychiatric injury, he could well serve a prohibition notice on an employer under section 22 of the Act.

Criminal prosecution based, for example, on a complaint from an employee of stress at work to an officer of the enforcing authority, or the failure by an employer to comply with an improvement notice or prohibition notice could be another line of enforcement.

9.7 Health and Safety (Display Screen Equipment) Regulations 1992

Both physical and mental stress are associated with work at display screens. The physically induced risks of visual and postural fatigue, and the potential for contracting a work-related upper limb disorder, are well recognized.

Work at display screens can be stressful in a number of other ways, however. For example, people may not have been trained properly in the use of the equipment, there may be insufficient space at the workstation, tasks may be inadequately designed, software may be badly designed or inadequate for the task, the system may not be easy to use or not provide appropriate feedback. In some organizations, individual performance may be monitored on a quantitative and qualitative basis, as a result of which employees may feel under pressure to meet output targets which have been set.

The principal requirements of the regulations are outlined below.

Regulation 2: Analysis of workstations to assess and reduce risks

1. Every employer shall make a suitable and sufficient analysis of those workstations which
   - (regardless of who has provided them) are used for the purposes of his undertaking;
   - for the purpose of assessing the health and safety risks to which those persons are exposed in consequence of that use.

Regulation 4: Daily work routine of users

Every employer shall so plan the activities of users in his undertaking that their daily work on display screen equipment is periodically interrupted by such breaks or changes of activity as to reduce their workload at that equipment.

Regulation 6: Provision of training

1. Where a person
   - is already a user on the date of coming into force of these Regulations; or

- is an employee who does not habitually use display screen equipment as a significant part of his normal work but is to become a user in the undertaking in which he is already employed,

his employer shall ensure that he is provided with adequate health and safety training in the use of any workstation upon which he may be required to work.

2. Every employer shall ensure that each user at work in his undertaking is provided with adequate health and safety training whenever the organization of any workstation in that undertaking upon which he may be required to work is substantially modified.

Regulation 7: Provision of information

1. Every employer shall ensure that operators and users at work in his undertaking are provided with adequate information about
   - all aspects of health and safety relating to their workstations; and
   - such measures taken by him in compliance with his duties under regulations 2 and 3 as relate to them and their work.
2. Every employer shall ensure that users at work in his undertaking are provided with adequate information about such measures taken by him in compliance with his duties under regulations 4 and 6(2) as relate to them and their work.
3. Every employer shall ensure that users employed by him are provided with adequate information about such measures taken by him in compliance with his duties under regulation 5 (eye and eyesight tests) and 6(1) as relate to them and their work.

9.7.1 The new technology

The introduction of new technology has, for many people, been stressful. For some older workers, who were brought up to use typewriters, for instance, coping with the new technology in terms of acquiring word processing skills, has required significant mental effort on their part. This is where the potential for stress can arise.

The provisions of the regulations apply to defined 'users' and 'operators' only and are quite straightforward. A user is defined as 'an employee who habitually uses display screen equipment as a significant part of his normal work'. An operator on the other hand, is 'a self-employed person who habitually uses display screen equipment as a significant part of his normal work'. It should be noted that the regulations do not apply to employees or self-employed persons using display screens who do not fit in with the above definitions.

9.7.2 Reducing stress in DSE users

An employer must implement a number of measures in order to reduce the risk of occupational stress amongst defined users and operators, including:

- analysis of workstations, which should include the potential for stress amongst new users and trainees;

- consideration of the daily work routine of users to permit breaks away from the screen;
- the provision of training, which should incorporate the manifestations of stress that can arise during display screen equipment work; and
- the provision of information on measures to be taken in the event of signs of stress-related ill health during or following display screen equipment work.

The schedule to the regulations sets out 'the minimum requirements for workstations'. These requirements, which are directed at either eliminating or reducing both the physical and mental stress of users and operators, must be taken into account when undertaking a workstation risk analysis (see below).

The schedule

(Which sets out the minimum requirements for workstations which are contained in Annex A to Council Directive 90/270/EEC on the minimum safety and health requirements for work with display screen equipment.)

Extent to which employers must ensure that workstations meet the requirements laid down in this schedule

1. An employer shall ensure that a workstation meets the requirements laid down in this schedule to the extent that:
   - those requirements relate to a component which is present in the workstation concerned;
   - those requirements have effect with a view to securing the health, safety and welfare of persons at work; and
   - the inherent characteristics of a given task make compliance with the requirements appropriate as respects the workstation concerned.

Equipment

2. (a) **General comment**

   The use as such of the equipment must not be a source of risk for operators or users.

   (b) **Display screen**

   The characters on the screen shall be well defined and clearly formed, of adequate size and with adequate spacing between the characters and lines.

   The image on the screen should be stable, with no flickering or other forms of instability.

   The brightness and the contrast between the characters and the background shall be easily adjustable by the operator or user, and also be easily adjustable to ambient conditions.

   The screen must swivel and tilt easily and freely to suit the needs of the operator or user.

   It shall be possible to use a separate base for the screen or an adjustable table.

   The screen shall be free of reflective glare and reflections liable to cause discomfort to the operator or user.

(c) **Keyboard**

The keyboard shall be tiltable and separate from the screen so as to allow the operator or user to find a comfortable working position avoiding fatigue in the arms or hands.

The space in front of the keyboard shall be sufficient to provide support for the hands and arms of the operator or user.

The keyboard shall have a matt surface to avoid reflecting glare.

The arrangement of the keyboard and the characteristics of the keys shall be such as to facilitate the use of the keyboard.

The symbols on the keys shall be adequately contrasted and legible from the design working position.

(d) **Work desk or work surface**

The work desk or work surface shall have a sufficiently large, low-reflectance surface and allow a flexible arrangement of the screen, keyboard, documents and related equipment.

The document holder shall be stable and adjustable and shall be positioned so as to minimize the need for uncomfortable head and eye movements.

There shall be adequate space for operators or users to find a comfortable position.

(e) **Work chair**

The work chair shall be stable and allow the operator or user easy freedom of movement and a comfortable position.

The seat shall be adjustable in height.

The seat back shall be adjustable in both height and tilt.

A footrest shall be made available to any operator or user.

Environment

3. (a) **Space requirements**

The workstation shall be dimensioned and designed so as to provide sufficient space for the operator or user to change position and vary movements.

(b) **Lighting**

Any room lighting or task lighting provided shall ensure satisfactory lighting conditions and an appropriate contrast between the screen and the background environment, taking into account the type of work and the vision requirements of the operator or user.

Possible disturbing glare and reflections on the screen or other equipment shall be prevented by co-ordinating workplace and workstation layout with the positioning and technical characteristics of the artificial light sources.

(c) **Reflections and glare**

Workstations shall be so designed that sources of light, such as windows and other openings, transparent or translucid walls, and brightly coloured fixtures or walls cause no direct glare and no distracting reflections on the screen.

Windows shall be fitted with a suitable system of adjustable covering to attenuate the daylight that falls on the workstation.

(d) **Noise**

Noise emitted by equipment belonging to any workstation shall be taken into account when a workstation is being equipped, with a view in particular to ensuring that attention is not distracted and speech is not disturbed.

(e) **Heat**

Equipment belonging to any workstation shall not produce excess heat which could cause discomfort to operators or users.

(f) **Radiation**

All radiation with the exception of the visible part of the electromagnetic spectrum shall be reduced to negligible levels from the point of view of the protection of operators' or users' health and safety.

(g) **Humidity**

An adequate level of humidity shall be established and maintained.

Interface between computer and operator/user

4. In designing, selecting, commissioning and modifying software, and in designing tasks using display screen equipment, the employer shall take into account the following principles:
   - software must be suitable for the task;
   - software must be easy to use and, where appropriate, adaptable to the level of knowledge or experience of the operator or user; no quantitative or qualitative checking facility may be used without the knowledge of operators or users;
   - systems must provide feedback to operators or users on the performance of those systems;
   - systems must display information in a format and at a pace which are adapted to operators or users;
   - the principles of software ergonomics must be applied, in particular to human data processing.

9.7.3 HSE guidance to the regulations

The HSE guidance to the regulations provides much advice on the best way to comply with the requirements of the regulations and the schedule. The sections of the guidance dealing with 'Task design and software' and 'Principles of software ergonomics' shown below are particularly significant in terms of the measures employers must take to prevent stress in users and operators of display screen equipment.

Task design and software

Principles of task design

Inappropriate task design can be among the causes of stress at work. Stress jeopardizes employee motivation, effectiveness and efficiency and in some cases it can lead to deterioration in health. The regulations are only applicable where health and safety rather than productivity is being put at risk. However, employers may find it useful to consider both aspects together as task design changes put into effect for productivity reasons may also benefit health, and vice versa.

In display screen work, good design of the task can be as important as the correct choice of equipment, furniture and working environment. It is advantageous to:

- design jobs in a way that offers users variety, opportunities to exercise discretion, opportunities for learning, and appropriate feedback, in preference to simple repetitive tasks whenever possible. For example, the work of a typist can be made less repetitive and stressful if an element of clerical work is added;
- match staffing levels to volumes of work, so that individual users are not subject to stress through being either overworked or underworked;
- allow users to participate in the planning, design and implementation of work tasks whenever possible.

Principles of software ergonomics

In most display screen work the software controls both the presentation of information on the screen and the ways in which the worker can manipulate the information. Thus software design can be an important element of task design. Software that is badly designed or inappropriate for the task will impede the efficient completion of the task and in some cases may cause sufficient stress to affect the health of a user. Involving a sample of users in the purchase or design of software can help to avoid problems.

Requirements of the organization and of display screen workers should be established as the basis for designing, selecting and modifying software. In many (though not all) applications the main points are:

- **Suitability for the task**: Software should enable workers to complete the task efficiently, without presenting unnecessary problems or obstacles.
- **Ease of use and adaptability**: Workers should be able to feel that they can master the system and use it effectively following the appropriate training.

  The dialogue between the system and the worker should be appropriate for the worker's ability.

  Where appropriate, software should enable workers to adapt the user interface to suit their ability level and preferences.

  The software should protect workers from the consequences of errors, for example, by providing appropriate warnings and information and by enabling 'lost' data to be recovered wherever practicable.
- **Feedback on system performance**: The system should provide appropriate feedback, which may include error messages; suitable assistance ('help') to workers on request; and messages about changes in the system such as malfunctions or overloading.

  Feedback messages should be presented at the right time and in an appropriate style and format. They should not contain unnecessary information.
- **Format and pace**: Speed of response to commands and instructions should be appropriate to the tasks and to workers' abilities.

  Characters, cursor movements and position changes should where possible be shown on the screen as soon as they are input.

- **Performance monitoring facilities**: Quantitative or qualitative checking facilities built into the software can lead to stress if they have adverse effects such as over-emphasis on output speed.

  It is possible to design monitoring systems that avoid these drawbacks and provide information that is helpful to workers as well as managers. However, in all cases workers should be kept informed about the introduction and operation of such systems.

9.8 The criminal implications of violence, bullying and harassment

The general duty of an employer to ensure, as far as is reasonably practicable, the health of his employees under section 2(1) of the HSWA applies where employees may be exposed to the risk of violence, bullying or harassment at work. Under section 2(2) there are extended duties:

- to provide systems of work that are, so far as is reasonably practicable, without risks to health; and
- to provide such information, instruction, training and supervision as is necessary to ensure, so far as is reasonably practicable, the health at work of his employees.

Violence at work has been the subject of considerable attention recently and employers need to be aware of employees at all levels who may be guilty of violent or aggressive behaviour whilst at work. To this extent, there should be a system for the confidential reporting to the employer of such behaviour towards themselves together with procedures for resolving these conflict situations. Many organizations have implemented a policy on violence at work which outlines their approach to this matter.

Failure by an employer to take this matter seriously could result in involvement with enforcement officers of the HSE or local authority. This could arise where an employee has, for example, made a formal complaint to the employer about aggressive behaviour, sex discrimination, racial abuse, harassment or physical violence towards himself by certain employees and the employer has taken no action. In this case it would not be unreasonable for the employee concerned to report this situation to the enforcement authority with a view to their taking action under the Health and Safety at Work etc. Act.

9.8.1 ACAS guidance for managers and employers: bullying and harassment at work

Everyone should be treated with dignity and respect at work. Bullying and harassment of any kind is in no-one's interest and should not be tolerated in the workplace. This guidance is designed to offer practical advice to employers to help them prevent bullying and harassment and to deal with any cases that occur. It includes guidelines for the development of policies and procedures.

What is bullying and harassment?

Examples and definitions of what may be considered bullying and harassment are provided below for guidance. For practical purposes those making a complaint usually define what they mean by bullying and harassment – something that has happened to them that is unwelcome, unwarranted and causes a detrimental effect. If employees complain that they are being bullied or harassed, then they have a grievance which must be dealt with regardless of whether or not their complaint accords with a standard definition.

How can bullying and harassment be recognized?

There are many definitions of bullying and harassment. Bullying can be characterized as offensive, intimidating, malicious or insulting behaviour, an abuse or misuse of power through means intended to undermine, humiliate, denigrate or injure the recipient.

Harassment, in general terms, is unwanted conduct affecting the dignity of men and women in the workplace. It may be related to age, sex, race, disability, religion, nationality or any personal characteristics of the individual, may be persistent or an isolated incident. The key is that the actions or comments are viewed as demeaning and unacceptable to the recipient.

Behaviour that is considered bullying by one person may be considered firm management by another. Most people will agree on extreme cases of bullying and harassment but it is sometimes the grey areas that cause most problems. It is good practice for employers to give examples of what is unacceptable behaviour in their organization and this may include:

- spreading malicious rumours, or insulting someone (particularly on the grounds of race, sex, disability, sexual orientation and religion or belief);
- copying memos that are critical about someone to others who do not need to know;
- ridiculing or demeaning someone – picking on them or setting them up to fail;
- exclusion or victimization;
- overbearing supervision or other misuse of power or position;
- making threats or comments about job security without foundation;
- deliberately undermining a competent worker by overloading and constant criticism;
- preventing individuals progressing by intentionally blocking promotion or training opportunities.

Bullying and harassment are not necessarily face-to-face. They may be written communications, electronic e-mail (so-called 'flame mail'), phone and automatic supervision methods, such as computer recording of downtime from work or the recording of telephone conversations, if these are not universally applied to all workers.

Bullying and harassment can often be hard to recognize. They may not be obvious to others and may be insidious. The recipient may think, 'Perhaps this is normal behaviour in this organization'. They may be anxious that others may consider them weak, or not up to the job, if they find the actions of others intimidating. They may be accused of 'over-reacting' and worry that they won't be believed if they do report incidents.

People being bullied or harassed may sometimes appear to over-react to something that seems relatively trivial but which may be the 'last straw' following a series of incidents. There is often fear of retribution if they do make a complaint. Colleagues may be reluctant to come forward as witnesses, as they too may fear the consequences for themselves. They may be so relieved not to be the subject of the bully themselves that they collude with the bully as a way of avoiding attention.

Why do employers need to take action on bullying and harassment?

Bullying and harassment are not only unacceptable on moral grounds but may, if unchecked or badly handled, create serious problems for an organization including:

- Poor morale and poor employee relations
- Loss of respect for managers and supervisors
- Poor performance
- Lost productivity
- Absence
- Resignations
- Damage to company reputation
- Tribunal and other court cases and payment of unlimited compensation.

It is in every employers' interests to promote a safe, healthy and fair environment in which people can work.

The 1991 European Commission Code *Protection of Dignity of Men and Women at Work* highlights the need for employers to develop and implement coherent policies to prevent harassment. In addition, various laws place responsibilities on employers to protect employees and these are outlined below.

The legal position

Discrimination and harassment

It is not possible to make a direct complaint to an employment tribunal about bullying. However, employees may be able to bring complaints under laws covering discrimination and harassment. For example:

- **Sex**: The Sex Discrimination Act gives protection against discrimination and victimization on the grounds of sex, marriage or because someone intends to undergo, is undergoing or has undergone gender reassignment.
- **Race**: The Race Relations Act 1976 gives protection against discrimination and victimization on the grounds of colour or nationality. The regulations that amended the Act (Race Regulations 2003) also give a stand-alone right of protection from harassment on the grounds of race and ethnic or national origin.
- **Disability**: The Disability Discrimination Act 1995 gives protection against discrimination and victimization.
- **Sexual orientation**: The Employment Equality (Sexual Orientation) Regulations 2003 give protection against discrimination and harassment on the grounds of sexual orientation. ('Orientation' is defined as 'same sex' – lesbian/gay; 'opposite sex' – heterosexual; and 'both sexes' – bisexual.)

- **Religion or belief**: The Employment Equality (Religion or Belief) Regulations 2003 give protection against discrimination and harassment on the grounds of religion or belief.

These regulations outlaw discrimination and harassment on grounds of religion or belief in all workplaces. They cover all aspects of the employment relationship, including recruitment, pay, working conditions, training, promotion, dismissal and references and enable employees to take prompt and effective action to tackle harassment.

As such, the regulations make illegal:

- Direct discrimination, that is, treating people less favourably because of their religion or belief.
- Indirect discrimination, that is, applying a provision, criterion or practice which disadvantages people of a particular religion or belief and which is not justified in objective terms.
- Harassment, namely unwanted conduct that violates people's dignity or creates an intimidating, hostile, degrading, humiliating or offensive environment.
- Victimization, namely treating people less favourably because of the action they have taken under, or in connection with, the legislation. An example could be victimization arising from an employee formally complaining of discrimination or giving evidence at a tribunal.

Unfair dismissal

Employers have a 'duty of care' for all their employees. If the mutual trust and confidence between employer and employee is broken, for example, through bullying and harassment at work, then an employee can resign and claim 'constructive dismissal' on the grounds of breach of contract. Employers are usually responsible in law for the acts of their workers.

Breach of contract may also include the failure to protect an employee's health and safety at work. Under the Health and Safety at Work etc. Act 1974 employers are responsible for the health, safety and welfare at work of all employees.

The Health and Safety Executive has recently addressed the issue of stress at work. *Guidance on Stress in the Workplace* includes the advice that 'stress should be treated like any other health hazard' and that employers have a 'legal duty to take reasonable care to ensure health is not placed at risk through excessive and sustained levels of stress'.

What should employers do about bullying and harassment?

First, they should consider framing a formal policy. This need not be over-elaborate, especially for small firms, and might be included in other personnel policies, but a checklist for a specific policy on bullying and harassment could include the following:

- statement of commitment from senior management;
- acknowledgement that bullying and harassment are problems for the organizations;
- clear statement that bullying and harassment will not be tolerated;
- examples of unacceptable behaviour;

- statement that bullying and harassment may be treated as disciplinary offences;
- the steps the organization takes to prevent bullying and harassment;
- responsibilities of supervisors and managers;
- confidentiality for any complaint;
- reference to grievance procedures (formal and informal), including timescales for action;
- investigation procedures, including timescales for action;
- reference to disciplinary procedures, including timescale for action, counselling and support availability;
- training for managers;
- protection from victimization;
- how the policy is to be implemented, reviewed and monitored.

The statement of policy will gain additional authority if staff are involved in its development. It should be made clear that the policy applies to staff on and off the premises, including those working away from base. The policy should also make plain that bullying or harassment of staff by visitors to the organization will not be tolerated.

All organizations, large and small, should have policies and procedures for dealing with grievance and disciplinary matters. Staff should know to whom they can turn if they have a work-related problem, and managers should be trained in all aspects of the organization's policies in this sensitive area.

Second, set a good example. The behaviour of employers and senior managers is as important as any formal policy. Strong management can unfortunately sometimes tip over into bullying behaviour. A culture where employees are consulted and problems discussed is less likely to encourage bullying and harassment than one where there is an authoritarian management style. The organization must make it clear that bullying and harassment are unacceptable.

Third, maintain fair procedures for dealing promptly with complaints from employees. Complaints of bullying and harassment can usually be dealt with using clear grievance and disciplinary procedures. Such procedures should have provision for confidentiality, and for the person making the complaint and the subject of the complaint to be accompanied by a fellow employee or trade union representative of their choice. (The right to be accompanied at grievance hearings is set out in the Employment Relations Act 1999.)

Fourth, set standards of behaviour. An organizational statement to all staff about the standards of behaviour expected can make it easier for all individuals to be aware of their responsibilities to others.

This may include information about what constitutes bullying and harassment. Many organizations find it helpful to supplement basic information with guidance booklets and training sessions or seminars. Training can also increase everyone's awareness of the damage bullying and harassment does both to the organization and to the individuals. Your staff handbook is also a good way of communicating with employees, and can include specific mention of the organization's views on bullying and harassment and their consequences.

Fifth, let employees know that complaints of bullying and/or harassment, or information from staff relating to such complaints, will be dealt with fairly and confidentially and sensitively. Employees will be reluctant to come forward if they feel they may be treated unsympathetically or are likely to be confronted aggressively by the person whose behaviour they are complaining about.

How should employers respond to a complaint of bullying and/or harassment?

Investigate the complaint promptly and objectively. Take the complaint seriously. Employees do not normally make serious accusations unless they feel seriously aggrieved. The investigation must be seen to be objective and independent. Decisions can then be made as to what action needs to be taken.

- **Informal approaches**: In some cases it may be possible to rectify matters informally. Sometimes people are not aware that their behaviour is unwelcome and an informal discussion can lead to greater understanding and an agreement that this behaviour will cease. It may be that the individual will choose to do this themselves, or they may need support from personnel, a manager, an employee representative or a counsellor.
- **Counselling**: In both large and small organizations counselling can play a vital role in complaints about bullying and harassment by providing a confidential avenue for an informal approach, and perhaps the opportunity to resolve the complaint without need for any further or formal action. Some organizations are able to train staff from within, others may contract with a specialist counselling service. Employee assistance programmes are counselling services provided and paid for by the employer and free to the employee. Business organizations may also be able to help in providing advice on accessing good counselling services. Counselling can be particularly useful where investigation shows no cause for disciplinary action or where doubt is cast on the validity of the complaint. Counselling may resolve the issue or help support the person accused as much as the complainant.
- **Discipline and grievances at work**: Where an informal resolution is not possible, the employer may decide that the matter is a disciplinary issue which needs to be dealt with formally at the appropriate level of the organization's disciplinary procedure. As with any disciplinary problem it is important to follow a fair procedure. In the case of a complaint of bullying or harassment there must be fairness to both the complainant and the person accused.

Guidance

Detailed guidance on how to handle disciplinary matters is available in the *ACAS Advisory Handbook: Discipline At Work*. The *ACAS Code of Practice: Disciplinary and Grievance Procedures in Employment* is reproduced in the handbook and provides advice on good practice in disciplinary matters which is taken into account in relevant cases appearing before employment tribunals. Briefly, a disciplinary procedure should:

- provide for matters to be dealt with quickly;
- ensure that individuals are made fully aware of what their disciplinary offence is;
- state the type of disciplinary action and who can take it;

- provide for a full investigation which gives individuals an opportunity to state their case;
- allow individuals to be accompanied by an employee representative or a colleague;
- not permit dismissal for a first offence (except for gross misconduct);
- ensure an explanation is given for disciplinary action; and
- specify an appeals procedure.

In cases which appear to involve serious misconduct, and there is reason to separate the parties, a short period of suspension of the alleged bully/harasser may need to be considered while the case is being investigated. This should be with pay unless the contract of employment provides for suspension without pay in such circumstances. Suspension without pay, or any long suspension with pay, should be exceptional as these in themselves may amount to disciplinary penalties. Do not transfer the person making the complaint unless they ask for such a measure.

There may be cases where somebody makes an unfounded allegation of bullying and/or harassment for malicious reasons. These cases should also be investigated and dealt with fairly and objectively under the disciplinary procedure.

What should be considered before imposing a penalty?

The action to be taken must be reasonable in the light of the facts. In some cases it may be concluded that a penalty is unnecessary or that counselling or training is preferable. The individual may now be more able to accept the need to change their behaviour. Where a penalty is to be imposed, all the circumstances should be considered including the employee's disciplinary and general record, whether the procedure points to the likely penalty, action taken in previous cases, any explanation and circumstances to be considered, and whether the penalty is reasonable.

Oral or written warnings, suspension or transfer of the bully/harasser are examples of disciplinary penalties that might be imposed in a proven case. Suspension or transfer (unless provided for in the employee's contract or agreed by the employer), could breach the employee's contract if they suffer detriment by it, for instance a transfer to a different location, which means additional expense, or a less responsible job. Any such breach could lead to a claim of constructive dismissal by the affected employee.

Where bullying or harassment amounts to gross misconduct, dismissal without notice may be appropriate.

Whenever a case of bullying or harassment arises, employers should take the opportunity to examine policies, procedures and working methods to see if they can be improved.

9.8.2 Indirect discrimination on the grounds of race or ethnic or national origin

Certain employees may be subject to indirect discrimination at work which can be stressful.

Indirect discrimination on the grounds of race or ethnic or national origin occurs when a person, A, applies to another person, B:

- a provision or criterion or practice which A applies to everyone; and
- the provision, criterion or practice puts (or would put) people from B's race or ethnic or national origin at a particular disadvantage; and
- the provision, criterion or practice puts B at a disadvantage; and
- A cannot show that the provision, criterion or practice is a proportionate means of achieving a legitimate aim.

An example of 'provisions, criteria or practice' which might be indirectly discriminatory could be a policy of an organization to ensure senior management positions are filled largely by existing managers who are white.

Colour or nationality

The definition of indirect discrimination in the Race Relations Act 1976 applies in complaints of discrimination based on grounds of colour or nationality. In these cases, indirect discrimination occurs when a person, A:

- applies to B a condition or requirement which A appears to equally apply to everyone; and
- the proportion of people from B's racial group (based on colour or nationality) that can comply is considerably smaller than the proportion of people not from that group who can comply with it; and
- A cannot justify the requirement or condition on non-racial grounds; and
- the requirement of condition is to B's detriment, because he or she cannot comply with it.

9.8.3 Harassment on racial grounds

Harassment of certain employees on racial grounds has been a feature of many workplaces in the past.

Under the Race Relations Act, this form of harassment is regarded as direct discrimination because it constitutes a 'detriment' in employment or in the way a service is provided. The Race Relations Regulations 2003 make harassment on grounds of race or ethnic or national origin a separate unlawful act. This will occur when a person, A, subjects another person, B, to unwanted conduct on grounds of race or ethnic or national origin that has the purpose or effect of:

- violating B's dignity; or
- creating an intimidating, hostile, degrading, humiliating or offensive environment for B.

Harassment on grounds of colour or nationality is classed, however, as possible direct discrimination under the Race Relations Act.

9.9 Whistle blowing

The term 'whistle blowing' has no legal definition within UK and EC law. However, it has been used to describe incidents where an employee publicly discloses some alleged wrongdoing within an organization.

What happens when an employee bypasses making a formal complaint to his employer of stress at work and goes direct to the enforcement agencies? Similar 'whistle blowing' activities could be carried out in respect of dangerous machinery or an unsafe working practice. In some cases, such complaints start off by means of an anonymous telephone call to the enforcing authorities.

This is a delicate situation for both the employer and the enforcement officer who has the task of investigating the complaint. Essentially, 'whistle blowing' protection was introduced in the Public Interest Disclosure Act 1999, which provides general protection for employees against dismissal or victimization by disclosing information on fraudulent, criminal or dangerous activities in the workplace. This Act aims to promote greater openness in the workplace and, by amending the Employment Protection Act 1996, protects 'whistle blowers' from detrimental treatment, that is, victimization or dismissal, for raising concerns about matters in the public interest. In providing this protection, the Act also reinforces the obligations of employees not to disclose to external sources any trade secrets or confidential information acquired during the course of their employment unless they fall within the disclosures qualifying for protection.

Employees who might be tempted to take such action should, however, realize that they do not necessarily have total immunity against action by their employer. In particular, an employee's misunderstanding or misinterpretation of the provisions of the Act could result in serious inconvenience for the employer and, in some cases, unwarranted attention by the enforcement agency. To this extent employees should be advised of the specific objectives of 'whistle blowing' so that they do not make unwarranted claims to the enforcement agencies or other official organizations.

9.9.1 Protected disclosure

Disclosure by an employee becomes a 'protected disclosure' only if it conforms to the strict requirements of the Act. The disclosure must be a 'qualified disclosure', that is, within the specified categories of wrongdoing or malpractice, which are made by a worker, employees or contract workers, including agency staff or home workers. Where these criteria apply, the employee is protected from any detrimental action by the employer. Detriment includes doing or not doing something to the worker that would amount to any kind of victimization. Furthermore, 'gagging' clauses in contracts are void in so far as they purport to preclude the worker from making a protected disclosure, including on termination of employment. Dismissal of an employee as a result of a protected disclosure is automatically considered as unfair, irrespective of length of service.

9.9.2 Qualifying categories

A 'qualifying disclosure' is restricted to any disclosure of information which, in the 'reasonable belief' of the worker making the disclosure, tends to show one or more of the following has happened, is happening or is likely to happen:

- a criminal offence;
- a person's failure to comply with any legal obligation to which he or she is subject;
- financial or non-financial maladministration or malpractice or impropriety or fraud;
- academic or professional malpractice;
- a miscarriage of justice;
- the endangerment of the health and safety of any person;
- damage to the environment;
- improper conduct or unethical behaviour; or
- deliberate concealment of information tending to show any of the above.

9.9.3 Policies on public disclosure

Many organizations, such as universities, financial institutions and those in the care industry, for example, hospitals, care homes and nursing homes now operate formal policies on public disclosure. Policies are designed to deal with concerns raised in relation to the specific issues which are in the public interest and listed above.

9.10 Home working and stress

Many people work at and from their home undertaking a range of activities of a professional, technical and manual nature. They include groups such as self-employed consultants, authors, managers, servicing technicians and those involved in activities such as packing and assembly of items. In recent years, many organizations have dispensed with the traditional office location and introduced the practice of 'teleworking' for their employees, very often as part of a wider set of flexible working options.

For many people, there are considerable benefits from working at home, such as flexible working hours, freedom from interruption, a degree of independence, no travelling costs to the office and back, no parking costs and the ability to look after their young children. Indeed, some people benefit from home working more than others in that they suffer less stress, are more self-motivated and feel they have greater control over their working life.

For others, the reverse situation is common, evidenced by feeling the pressure to work longer hours, the stress associated with balancing family demands with those of the employer and the 'invasion' of their home life and space.

On this basis, home working can be a 'double-edged sword' in terms of the disadvantages and advantages that it creates. Much will depend upon the home circumstances of the employee and whether home working suits these circumstances at that stage in their lives in terms of, for example, space availability, family commitments, time flexibility and the costs involved.

Whilst home working can offer considerable benefits to both employers and employees, it is important that the organization has some form of policy to deal with this aspect of working. Health and safety issues feature prominently in such a policy, particularly with respect to the prevention and control of stress amongst home workers. In many organizations, potential home workers are asked to complete a lengthy risk assessment of their homes, in particular the area to be used for working. There is the question of office furniture and equipment to be provided, along with, for instance, personal computers, printers, telephones and telephone lines. It is common for the home to be inspected by a safety adviser prior to the commencement of home working and for electrical installations to be checked by a competent person. The HSE (2003) publication *Home Working: Guidance for Employers and Employees on Health and Safety* provides useful guidance on this issue.

Whilst the physical environment can be easily controlled, the potential for stress amongst home workers must be considered. For example, the International Labour Organization draws attention to the risk of 'workaholism' among teleworkers, with teleworkers more likely than office staff to find work demands increasingly conflicting with their home lives. In fact, for some people, there may be a serious problem in achieving the physical separation necessary and the appropriate boundaries between family life and that of their work. In some cases, the work can completely take over the individual's home life, characterized by, for example, the dining table being loaded with documents, a personal computer and other items. In other cases, the reverse takes place, whereby home concerns risk swamping out the work.

The isolation associated with home working is another factor which, for some, is stressful. Many home workers miss the social aspects of work, including regular contact with colleagues and participation in discussions about work. One remedy, practised in some organizations, is for all home workers to attend at the workplace for, say, one day a week to attend meetings, receive new work and to discuss any problems with their manager. This reduces the feeling of isolation and, for the employer, achieves 'the best of both worlds'.

9.11 Disability Discrimination Act 1995

The Disability Discrimination Act (DDA) makes it unlawful for an employer to discriminate against a disabled job applicant or worker with respect to:

- Selection for jobs
- Terms and conditions of employment
- Promotion or transfer
- Training
- Employment benefits
- Dismissal or other detrimental treatment.

The Act does not apply to prison officers, fire fighters, police officers, members of the armed forces, people working on board ships, aircraft and hovercraft. It further does not apply to employees working wholly or mainly abroad or to employers who employ less than 15 employees.

The Act does cover permanent, temporary and contract workers, as well as full-time employees. It applies to employees or potential employees with a 'physical or mental impairment', which has long-term effects on their ability to carry out normal day-to-day tasks. Part II of the Act deals with discrimination with respect to employment.

9.11.1 Definitions of 'disability' and 'disabled person'

A person has a disability for the purposes of the Act if he has a physical or mental impairment which has a substantial and long-term adverse effect on his ability to carry out normal day-to-day activities. A 'disabled person' means a person who has a disability.

9.11.2 Physical or mental impairment

The term 'impairment' is not defined in the Act. However, in the guidance accompanying the DDA reference is made to the matters to be taken into account in relation to the discrimination of disability. This guidance expands on the definition of 'mental impairment' as not including any impairment resulting from a mental illness unless that illness is 'clinically well-recognized'.

The Disability and Discrimination (Meaning of Disability) Regulations 1996 list conditions which would not be treated as amounting to an impairment, such as addictions to alcohol, nicotine or any other substance unless originally caused by taking prescribed drugs.

9.11.3 Substantial and long-term effects

'Long-term' means at least 12 months or the remainder of the person's life (if less than 12 months). Where a condition is recurrent, it is treated as continuing if it is likely to recur. This would apply to a number of conditions, such as epilepsy. Indicators of potentially substantial effects are the time taken to undertake tasks and the way in which tasks are carried out.

Consideration must be given to the cumulative effects of impairments which may not in themselves be substantial if taken individually. The effect of environmental conditions, such as temperature, humidity or fatigue are relevant to the issue of whether or not the adverse effect is 'substantial'.

9.11.4 Modifying behaviour

The extent to which an employee can reduce the effects of his condition by modifying his behaviour must be taken into account, though the beneficial effects of treatment must be disregarded when assessing the severity of these effects. For example, the degree of impairment of a person using a hearing aid would be assessed by the level of hearing without the use of the hearing aid. Similar provision applies in the case of a person with an artificial limb, but not in the case of people wearing spectacles or using contact lenses.

9.11.5 Severe disfigurement

This is classed as a 'substantial adverse effect' unless it is a disfigurement arising from a tattoo or body piercing.

9.11.6 Normal day-to-day activities

A person is defined as disabled only if his impairment affects his ability to carry out one of the following day-to-day activities:

- Mobility
- Manual dexterity
- Physical co-ordination
- Continence
- Ability to lift, carry or otherwise move everyday objects
- Speech, hearing, eyesight
- Memory or ability to concentrate, learn or understand
- Perception of risk of physical danger.

The guidance lists examples of effects which would amount to an impairment and those which would not do so. For example, a person who can walk no more than a mile without discomfort would have difficulty establishing this inability as an impairment. In the case of a person's ability to lift, the guidance indicates that the inability to pick up objects of moderate weight with one hand would be a substantial adverse effect, whereas the inability to move heavy objects without some form of mechanical aid would not be so.

9.11.7 Past and future disability

The Act applies to persons who had a disability in the past but not to persons with a genetic predisposition or future risk of disability.

9.11.8 Disability discrimination

There are two main elements to disability discrimination, namely:

- a duty not to discriminate for a reason relating to a disability; and
- a duty to adjust.

Section 5(1) specifies two elements to test whether an employer has discriminated against a disabled person. An employer will have discriminated if:

- for a reason which relates to the disabled person's disability, he treats him less favourably than he treats or would treat others to whom that reason does or would not apply;
- he cannot show that the treatment in question is justified.

The term 'relates' above can be used to challenge what has been known in racial discrimination and sex discrimination cases as 'indirect discrimination'. This means that a disabled person who is treated less favourably as a result of an employer's requirement or policy and which is to the disadvantage of the disabled person may be protected by the DDA.

On this basis employers need to exercise care in the selection of job applicants and in the specification of job requirements. For example, a requirement that an office worker should have a minimum level of computer literacy or that a machine operator should be capable of handling certain loads, may be discriminatory against a disabled person in a number of ways. For instance, that person may have some form of learning difficulty or be physically handicapped.

9.11.9 Justification: an employer's defence

Where it is alleged at an employment tribunal that an employer discriminated against a disabled person, the employer may submit a defence that the discriminatory treatment is 'justified'. Discrimination will only be justified, however, if 'the reason for it is both material to the circumstances of the particular case and substantial'. The burden to prove justification rests with the employer. What is 'material' and 'substantial' rests with the tribunal. However, decided cases indicate it is a low threshold for employers to overcome.

The Code of Practice accompanying the DDA provides guidance for employers, employees and tribunals. It indicates that employee or customer preference will not usually justify discrimination and further warns against unjustifiable medical examinations and fitness requirements.

9.11.10 The employer's duty to adjust

Employers must make reasonable adjustments where working arrangements and/or the physical features of a workplace cause a substantial disadvantage for a disabled person in comparison with those who are not disabled.

Reasonable steps which employers may need to take include:

- Altering working hours
- Allowing time off for rehabilitation or treatment
- Allocating some of the disabled person's duties to someone else
- Transferring the disabled person to another vacancy or another place of work

- Giving or arranging training
- Providing a reader or interpreter
- Acquiring or modifying equipment or reference manuals
- Adjusting the premises.

Failure to make a 'reasonable adjustment' is a separate act of discrimination unless it can be justified. Furthermore, an employer cannot rely on a failure to make a reasonable adjustment in order to establish a justification defence for the duty not to discriminate under section 5(1). So where something which might amount to a material and substantial reason for treating a disabled person less favourably could be made insubstantial by a reasonable adjustment, an employer would not be successful in a justification defence.

9.11.11 Arrangements

Section 6(2) provides that an employer is under a duty to make a reasonable adjustment to:

- Arrangements to determine who is offered employment, i.e. selection arrangements
- Terms and conditions or other arrangements on which employment, promotions, transfer, training or any other benefit is offered.

9.11.12 Physical features of workplaces

Physical features are:

- those arising from the design or construction of a building;
- exits or access to buildings;
- fixtures, fittings, furnishings, equipment or materials;
- any other physical element or quality of land or the premises.

Typical examples of physical features that might cause substantial disadvantage to a disabled person include the absence of ramps for wheelchair users, inadequate lighting for someone with restricted vision, doors that are too narrow for wheelchair users and a chair which is unsuitable.

9.11.13 Reasonable steps

In considering what is a reasonable step for an employer, reference can be made to the cost of the adjustments, their effectiveness and the financial resources of the employer.

9.11.14 Victimization

Where a person is involved in bringing proceedings, making allegations or giving information in connection with the DDA, he is protected against victimization. This means that he must not be subject to any 'detriment' as a result.

9.11.15 Complaints to an employment tribunal

A complaint of discrimination can be taken to an employment tribunal. Procedures and remedies available are similar to those under the Sex Discrimination Act 1975 and the Race Relations Act 1976. In particular,

- There is a 3-month time limit for lodging tribunal applications.
- An employment tribunal has the power to make a declaration, recommendation or to award compensation on which interest can be payable.
- There is provision for the use of a questionnaire procedure in the case of employers.

9.11.16 Code of practice

The Act is accompanied by a code of practice for the elimination of disability discrimination. Whilst the code of practice does not impose duties on employers, it can be used in evidence before a tribunal and the recommendations of the code of practice must be taken into account.

The code of practice includes many examples of what amounts to unlawful discrimination.

9.11.17 Other aspects of disability discrimination

Part III of the DDA deals with discrimination in other areas, such as goods, facilities and services, and in respect of premises. Part IV covers disability discrimination in education and Part V with respect to public transport. Part VI makes provision for the establishment of the National Disability Council.

9.12 Disability arising from mental impairment

The question of disability arising from 'mental impairment' has recently been raised. In considering the definition of stress, an Employment Appeal Tribunal (www.workplacelaw.net 30.1.2002) ruled that, in order to decide whether someone had a 'mental impairment' causing disability, medical evidence must show either:

1. that the applicant was suffering from a clinically well-recognized mental illness; or
2. that the applicant had some form of mental impairment, which does not arise from a mental illness at all, but from some other causes.

It should be pointed out that this distinction, which arises from one special interpretation of the Act, is important as, in the past, mental impairment caused by anxiety, depression, stress and other related symptoms, have been excluded as defined 'disabilities' under the Act.

On this basis, an employer can be liable under the DDA for treating an employee less favourably on the grounds of disability if that disability is depressive illness resulting from stress at work.

Claims under the discrimination legislation must be lodged with an employment tribunal within 3 months of the discriminatory act. There is no statutory limit on the compensation which can be awarded. Compensation can also include that for 'injury to feelings', and for any personal injury sustained by an employee as a consequence of the discriminatory treatment.

9.13 Employment tribunals and stress

In some cases, an employee may be so stressed by aspects of his work that he may decide to resign. An alternative course of action, as opposed to an action for negligence in the civil courts, is an action for unfair dismissal or discrimination before an employment tribunal.

9.13.1 The Employment Rights Act 1996 and bullying

Unfair dismissal

Under section 94 of the Employment Rights Act (ERA) 1996 every employee has the right not to be unfairly dismissed by his employer. In order to make an application for unfair dismissal, an employee must have continuous employment with the same employer for at least one year prior to the effective date of termination of employment, and must have been dismissed. The Act recognizes two further acts by an employer which constitute dismissal, namely the non-renewal of a temporary contract and 'constructive dismissal' where, for example, an employee becomes so stressed by particular aspects of his work or working environment, that he resigns.

Section 95 of the Act states:

An employee shall be treated as dismissed by an employer if:

(a) the contract under which he is employed is terminated by the employer, with or without notice;

(b) he is employed under a contract for a fixed term, that term expires without being renewed under the same contract; or

(c) the employee terminates the contract under which he is employed, with or without notice in circumstances in which he is entitled to terminate it without notice by reason of the employer's conduct.

Constructive dismissal

Constructive dismissal is largely based on the third reason above. However, in an action for constructive dismissal it is for the employee to prove that the employer has committed a fundamental breach of the contract of employment by reason of the employer's conduct.

In a case where an employee, who has sufficient continuity of employment, applies to an employment tribunal then, under section 98 of the ERA, it is for the employer to show the reason for the dismissal, and that it is a reason which relates to capability or qualifications, conduct, redundancy, breach of a statute or some other reason which is significant enough to justify dismissal. 'Capability' with respect to an employee means his capability assessed by reference to skill, aptitude, health or any other physical or mental quality. The question of stress and stress-related ill health arises in this case.

Furthermore, under section 98 of the ERA, the determination of the question as to whether dismissal was fair or unfair, having regard to the reason(s) shown by the employer, depends upon whether in the circumstances, including the size and administrative resources of the employer's undertaking, the employer acted reasonably or unreasonably in treating it as a sufficient reason for dismissing the employee, and that question shall be determined in accordance with equity and the substantial merits of the case.

Sources of stress at work include bullying, discrimination, underpay and overwork. Other sources of employment legislation are the Sex Discrimination Act 1975, the Race Relations Act 1976 and the DDA, together with the National Minimum Wage Act 1998 and the Working Time Regulations 1998. Breaches of this legislation by an employer could be a source of stress to employees. In the case of the DDA a person has a disability if he has a physical or mental impairment which has a substantial long-term effect on his ability to carry out normal day-to-day activities. Stress or stress-related illness, caused by long-term depression, is common. Depression is a condition covered by the DDA and there is no limit to the damages that can be awarded for this condition.

In *Ward v Signs by Morrell Ltd*, which was one of the first cases of disability discrimination brought before a tribunal, the criteria laid down in the Act, which the applicant needed to satisfy in order to succeed in the action for such discrimination, were specified, namely:

- The applicant must have a mental impairment, that is, an impairment clinically well recognized by a respected body of medical opinion.
- The impairment must have adverse effects which are substantial.
- The substantial effect must be long-term.
- The long-term substantial effects must have an adverse effect on normal day-to-day activities.

Stress caused by excessive working hours is controlled by the maximum working time imposed by the Working Time Regulations 1998.

Furthermore, employment tribunals can now hear breach of contract claims where damages are limited to £25 000. What is significant is that it is less costly to bring a case at an employment tribunal, as currently necessary, than to seek redress in the civil courts.

Discrimination and unfair dismissal

In cases of discrimination it is for the individual employee to bring an action for discrimination against the employer before an employment tribunal. Under the DDA, a person has a disability for the purposes of the Act if 'he has a physical or mental impairment which has a substantial and long-term adverse effect on his ability to carry out normal day-to-day activities'. Whether such a definition includes stress, or stress-related illness, has yet to be fully decided by the courts and tribunals.

9.13.2 Working Time Regulations 1998

These regulations implemented the provisions of the European Working Time Directive and had something of a chequered introduction to the UK largely because the European legislators intended the directive to be the arbiter of safe practice in the workplace. However, the UK Government perceived the directive to be more a social measure because it sought to interfere with the traditional working patterns in the UK. However, this argument was rejected and the government was obliged to implement the directive as a health and safety measure.

UK workers tend to work longer hours than workers in other European countries, a practice which is unlikely to change. Stress and the greater potential for accidents are two outcomes of inadequate sleep, this correlation between sleep loss and reduced or impaired mental performance being confirmed by a range of studies on this aspect. Classic examples of this correlation can be seen in the Chernobyl incident, the Challenger shuttle explosion and the Exon Valdez oil tanker disaster. In all three cases, situations arose where employees made inappropriate decisions, or none at all, due to their sleep-deprived state.

Weekly working time

One of the main areas of contention in the regulations is the provision that weekly working time, including any overtime working, is limited to a maximum of 48 hours over a standard averaging time of 17 weeks. However, employees can enter into a written agreement with their employers to exclude that requirement. This 48 hour limit can raise significant problems for certain industries who rely on overtime in order to ensure the completion of projects on time, as in the case of the construction industry and certain areas of manufacturing.

Night work

In the case of night work, employers are required to take all reasonable steps to ensure that the normal hours of their night workers do not exceed an average of 8 hours in each 24 hour period over a 17 week averaging period. Where a night worker's work involves special hazards or heavy physical or mental strain, there is a limit of 8 hours on the worker's actual daily working time.

Health assessment

Night workers are entitled to a free health assessment before taking on night work and at regular intervals afterwards. The aim of the health assessment is to determine whether the employee is fit to undertake the night work to which he has been assigned. There is no formal procedure for undertaking a health assessment. According to the guidance produced by the Department of Trade and Industry, as a minimum employers should compile a screening questionnaire for employees to complete before commencing night work.

Such a questionnaire, compiled by an occupational physician or occupational health nurse, should seek evidence of a number of medical conditions that could arise or be aggravated by night work, such as diabetes, cardiovascular, gastrointestinal and sleep disorders. The questionnaire should be completed on an annual basis thereafter. Responses to the questionnaire should be evaluated by health professionals with particular reference to the potential for injury to health arising from sleep deprivation resulting in stress in the employees concerned.

9.14 HSE research reports

A recent HSE research report entitled *An Assessment of Employee Assistance and Workplace Counselling Programmes in British Organizations* outlined some of the remedies for employers thus:

> *Occupational stress is a feature of the current economic climate and most people suffer from it at times and to different degrees. Occupational stress can seriously impair the quality of work life and reduce personal and job effectiveness. It can result from the job itself and from the context and arrangement of work but, equally, it can originate from outside factors such as personal or family life. Whatever the origins, there is no doubt that work can be stressful.*

Thus, the report indicates, stress management programmes have the potential to promote an employee's positive contribution and wellness at work and, as a consequence, to encourage organizational learning. In this respect, stress management programmes have some compatibility with corporate strategy, as well as the ability to become incorporated into the organization's attitudes, discourses and practices. A stress management programme can thus become part of corporate culture and make both quantifiable and qualitative contribution to organizational effectiveness.

According to the HSE publication *Stress at Work: A Guide for Employers* (1995) the benefits of stress management programmes include better health for employees, reduced sickness absence, increased performance and output, better relationships with clients and colleagues and reduced staff turnover.

Other HSE research reports on stress prevention and rehabilitation are of particular significance.

- *Beacons of Excellence in Stress Prevention* (2003) outlines criteria for best practice in stress prevention. These criteria were developed from a literature review of recent research in this area and through expert opinion. The criteria were used to identify organizations that could be considered examples of best practice in various aspects of stress prevention.

- *Best Practice in Rehabilitating Employees Following Absence due to Work-related Stress* (2003) clarifies criteria of current best practice in rehabilitation. The report provides clear and practical steps that employers can put in place to encourage employees to return to work and prevent a recurrence of the initial stress. The case studies in this report cover England and Wales. One section provides specific advice for small- and medium-sized companies.

9.15 HSE management standards

The case studies outlined in the above research reports were used, in conjunction with other examples of good practice gathered by the HSE's Stress Priority Programme Team, to develop guidance to support employers in implementing the HSE's management standards (see Chapter 7) by outlining practical solutions to assist them to tackle stress at work.

9.16 Conclusion

The increasing number of civil cases arising from stress-induced injury foreshadows the potential for the government of the day to produce legislation in written form. This could, perhaps, be by means of a statute, regulations made under the HSWA or an approved code of practice under the Management of Health and Safety at Work Regulations dealing with the subject of stress at work.

Whilst stress is not specifically referred to in existing regulations, such as the Management of Health and Safety at Work Regulations, the potential for psychiatric injury must be taken into account in the risk assessment process under these regulations. Where such injury is likely, management systems must be put into operation to either prevent or control exposure to stress. This includes the need to consider human capability when allocating tasks and to provide information which is comprehensible and relevant in this respect to employees.

The general duty on employers to protect, so far as is reasonably practicable, the 'health' of their employees under the HSWA is significant within the context of the potential for psychiatric injury. It is conceivable that pressure groups, such as trade unions, will inevitably bring pressure on the enforcement agencies, such as the Health and Safety Executive, to test this matter in the criminal courts through a test prosecution of an employer in the not too distant future.

On this basis, employers need to take this matter in hand with a view to producing appropriate strategies to prevent or control stress that meet the general requirements of current criminal health and safety and employment legislation.

Questions to ask yourself after reading this chapter

- Does the organization's statement of health and safety policy under the Health and Safety at Work etc. Act make specific reference to the measures necessary to reduce or control stress amongst employees?
- Do you have a separate stress at work policy?
- Have you considered stress on employees when undertaking a risk assessment under the Management of Health and Safety at Work Regulations?
- What information about stress have you given to your employees?
- Have you undertaken workstation risk analyses for defined display screen equipment 'users' in accordance with the Health and Safety (Display Screen Equipment) Regulations 1992?
- Do you undertake undisclosed monitoring of display screen equipment users?
- Have you considered the ergonomic aspects of display screen equipment workstation design?
- What measures are in place to protect employees against violence, bullying and harassment at work?
- What arrangements have been made to ensure home workers are not exposed to excessive stress?
- Is there evidence of disability discrimination in the organization?
- What measures have been taken to comply with the requirements of the Disability Discrimination Act?
- Do your employees suffer sleep deprivation due to the pattern of hours they have to work?

Key points – implications for employers

- The criminal law implications of stress at work is a matter which enforcement agencies are now having to consider.

- The criminal implications of stress hinge around:
  (a) the general duty of employers under section 2 of the Health and Safety at Work etc. Act to ensure, so far as is reasonably practicable, the health of all persons at work; and
  (b) duties under the Management of Health and Safety at Work Regulations with respect to risk assessment, the development and implementation of management systems and the consideration of human capability when allocating tasks.

- The use of display screen equipment can be stressful for some people. Employers should undertake display screen equipment workstation risk assessments for identified 'users' and implement any recommendations arising from same in accordance with the Schedule to the Health and Safety (Display Screen Equipment) Regulations 1992.

- Disability arising from 'mental impairment' has recently been considered by the courts under the Disability Discrimination Act.
- Stress may feature in claims for unfair and constructive dismissal before an employment appeals tribunal in some cases.

Appendix: Stress audit

This audit is designed to enable employers:

- to identify sources of stress in the workplace;
- to establish which groups of employees are more at risk;
- to evaluate the severity of the risk;
- to assess and evaluate current strategies for dealing with stress at work;
- to specify the further measures that may be necessary over a period of time to reduce stress amongst all groups of employees; and
- to review the situation at a later date.

The audit takes the form of a series of questions directed at ascertaining the existing level of stress and future potential for stress amongst employees. It incorporates an Action Plan which should identify short-, medium- and long-term recommendations for action.

Undertaking the audit

The audit should preferably be undertaken by a designated group, such as a senior manager, health and safety specialist or occupational health nurse, who has been trained in stress management, and employee representatives. All the questions should be considered carefully.

Elements of the HSE's Management Standards for Stress have been incorporated into the audit.

What is stress?

For the purposes of this stress audit, 'stress' is defined as 'pressure and demands placed on a person beyond his ability to cope' (HSE, 1995).

1. Current information
   - What is the current level of sickness absence?
   - Are the causes of sickness absence classified?
   - Is the current level of sickness absence satisfactory compared with other organizations in the particular industry or business group?
   - Are managers aware of those on sick leave who may be suffering the effects of work-related stress?
   - Is current time-keeping satisfactory?
   - Are certain employees habitually late for work?
   - Is there evidence of increased alcohol consumption amongst some employees?
   - Do some employees complain of sleep and digestive disorders?
   - Have all employees completed a Personal Stress Questionnaire?
   - Have the results of the Personal Stress Questionnaire been analysed with a view to identifying the principal causes of stress amongst employees?

2. The organization
   - Are there signs of stress amongst employees?
   - Are there problems arising from stress in the organization?
   - Do managers appreciate that their attitudes, approaches, ability to make decisions and the standards they set may result in stress?
   - Is the organization well run or is it a case of 'management by crisis' all the time?
   - Are managers trained to recognize the manifestations of stress in individuals?
   - Are attempts made to ensure consistency of management style and approach?
   - Do some managers have difficulty in delegating responsibility?
   - Are employees occasionally subjected to rumours of redundancy, downsizing or other causes of job loss?
   - Do employees receive regular feedback on the success or otherwise of the organization's operations and activities?
   - Do the majority of employees indicate that they understand their role and responsibilities?
   - Are there restrictions on the way people should behave?
   - Do some people have responsibility for other people's lives?

3. Work activities
   - Can the majority of employees cope with the demands of their jobs?
   - Are the majority of employees sufficiently competent to undertake the core functions of their jobs?
   - Are job descriptions clear and unambiguous?
   - Are many jobs repetitious, boring and uninteresting?
   - Are job characteristics considered by management with respect to the potential for stress?
   - Do risk assessments consider the potential for stress and its effects on employees?
   - Do some jobs cause high levels of physical and mental fatigue?
   - Do some jobs result in loss of motivation amongst employees?

- Is the error rate in some jobs, or with certain individuals, higher than normal?
- Are effective measures taken to reduce operator error?
- Are the causes of certain accidents attributable to stress?
- Do employees have a high degree of control and input over how they do their work?
- Are employees encouraged to develop new skills?
- Are employees provided with adequate support when asked to undertake new tasks?
- Are employees able to communicate with each other satisfactorily?
- Are employees put under pressure to meet deadlines on a regular basis?
- Do some tasks involve complex decisions, diagnoses or calculations?
- Are employees expected to work overtime regularly?
- Does the organization operate a shiftwork system?
- Are flexible working hours available for part-time employees?
- Do some employees have to undertake tasks for which they may be inadequately trained?
- Do some employees have to deal with members of the public on a regular basis?
- Is there evidence of sexual and/or racial harassment of certain employees by managers, other employees, customers or members of the public?
- Are employees encouraged to report evidence of such harassment?
- Is co-ordination between departments satisfactory?
- Is there flexibility in work procedures?
- Are ergonomic factors considered in the design of jobs?
- Are machinery and equipment well maintained?
- Do night workers complain of isolation from their families?

4. The workplace
    - Is the workplace well-designed, adequately heated, lit and ventilated?
    - Are workplace noise levels excessive?
    - Do some people work in isolated locations?
    - Is there sufficient space for employees to operate comfortably, safely and in the most efficient manner?
    - Do employees have sufficient privacy, without distraction, to enable them to concentrate?
    - Are working areas cleaned regularly and effectively?
    - Are some employees exposed to chemical and/or biological stressors?
    - Does the nature of the work expose employees to the risk of contracting physically induced conditions, such as work-related upper limb disorders?

5. Documentation
    - Does the organization have a Statement of Policy on Stress at Work?
    - Is this Statement of Policy regularly reviewed in the light of experience?
    - Are Work-Related Stress Risk Assessments undertaken, recorded and the recommendations from same implemented?

6. The organizational culture
   - Could the organization's culture be described as 'friendly' or otherwise?
   - Do people actually enjoy coming to work for the organization?
   - Is the mental and physical capability of employees taken into account when allocating tasks?
   - Does the organization recognize and reward high levels of performance?
   - Are the majority of employees resistant to change?
   - Is there emphasis on competitiveness?
   - Do some employees experience conflict between work and outside interests?
   - Do the majority of employees indicate that the organization engages them frequently when undergoing an organizational change?
   - Does the organization ensure that the employees understand the reason for proposed changes?
   - Do employees receive adequate communication during the change process through activities such as team briefing, management meetings, in-house broadsheets, one-to-one communication, etc?
   - Are employees made aware of the impact of the changes on their jobs?
   - Do employees receive adequate support during the change process?
   - Does the organization endeavour to enforce its standards with respect to behaviour, dress, behavioural norms and standards?
   - Do managers and employees find job and career reviews/appraisals stressful?
   - Do employees have to work long hours?
   - Does the organization run a suggestion scheme?
   - Do employees feel resentful when their suggestions to improve work procedures are disregarded by management?
   - Is there evidence of a culture of bullying and/or harassment?

7. Relationships between employees
   - Do some employees have a reputation for aggressive and offensive behaviour towards other employees?
   - Do the majority of employees indicate that they are not subjected to unacceptable behaviour, such as bullying?
   - Are complaints of bullying, harassment or other forms of conflict investigated thoroughly by management with a view to effective prevention or quick resolution?
   - Is there a confidential procedure for reporting violence of all types at work?
   - Is there a formal Statement of Policy on Violence at Work?

8. Employee support
   - Is there a formal procedure to enable employees to report stressful conditions?
   - Are employees encouraged to seek support at an early stage if they feel as though they are unable to cope?
   - Are such employees provided with support and advice from an occupational health practitioner when experiencing stress arising from work or their home/family circumstances?

9. Information, instruction and training
   - Do employees receive adequate information, instruction and training:
     (a) on induction?
     (b) on being exposed to new or increased risks arising from
       (a) transfer or being given a change in responsibility?
       (b) the introduction of new work equipment or a change in work equipment?
       (c) the introduction of new technology?
       (d) the introduction of a new system of work or change in an existing system of work?
   - Is information provided comprehensible and relevant?
   - Do employees receive information, instruction and training on:
     (a) the causes and effects of stress?
     (b) the reasons for and methods of introducing change?
     (c) the need to maintain a healthy lifestyle?
     (d) time management?
     (e) relaxation?
     (f) dealing with personal crisis?
     (g) personal assertiveness?
     (h) interpersonal skills?
     (i) Does the organization run stress management training for employees?

10. Health surveillance
   - Are employees subject to regular health surveillance by an occupational physician or occupational health nurse?
   - Is stress counselling available to employees diagnosed as suffering from stress?
   - Are health promotion arrangements adequate?
   - Are the causes of sickness absence discussed with individual employees?
   - Is sufficient attention paid to women's health issues, such as ensuring a healthy pregnancy and the menopause?
   - Are employees advised on lifestyle factors such as smoking cessation, healthy eating, weight control, personal hygiene measures and the need for physical activity?

Action plan

Short-term action

Medium-term action

Long-term action

Date

Audit group members

Date of audit review (12 months)

10

Executive summary

1. The HSE define stress as 'pressure and extra demands placed on a person beyond his ability to cope'.
2. It is important to distinguish between positive stress, or 'pressure', as opposed to negative stress which can result in psychiatric injury.
3. Stress at work is a significant hidden cost to employers.
4. Stress-related ill health is now recognized by the courts and, in recent years, there has been a 90 per cent increase in civil claims for mental and physiological damage.
5. Stress is associated with the autonomic system, the body's 'flight or fight' response.
6. Certain occupational groups, such as teachers and police officers, are more likely to suffer stress.
7. There are many direct and indirect causes of stress at work.
8. Atypical workers, such as shift workers and night workers, are particularly prone to stress.
9. Violence, bullying and harassment of employees at work has been brought to light in recent civil claims.
10. The HSE have produced recommendations on violence management at work.
11. Symptoms of stress vary dramatically from person to person and can be of a long- and short-term nature.
12. Stress can have direct effects on job performance.
13. Anxiety and depression are classic manifestations of stress.
14. Personality traits have a direct relationship with stress in individuals.
15. Employers need to recognize that many employees go through some form of personal crisis at some time in their lives.
16. The problems of smoking, alcohol abuse and drug addiction at work must be addressed.
17. Women at work can be subject to many stressors not suffered by their male counterparts.

18. There are techniques available for the measurement and evaluation of stress.
19. People need training in order to be able to respond satisfactorily to stressful events in their lives.
20. Change in organizations is one of the most prominent causes of stress.
21. People may need time management and assertiveness training to enable them to cope better with stress at work.
22. Relaxation therapy is an excellent means of reducing stress.
23. HSE guidance on stress in the workplace includes a range of strategies for employers for preventing or reducing stress; namely good management, the right attitudes to stress by employers, management style and procedures for dealing with change, together with general management and culture, relationships at work, work schedules, decision-making and policy, clear definition of the employee's role, job design and workload or workplace.
24. Employers need to recognize evidence of stress in the workplace evidenced by standards of work performance, staff attitude and behaviour, relationships at work and sickness absence levels.
25. Employers need to pay greater attention to the human factors in elements of work with particular reference to the attitudes of employees, motivational factors and personality traits as they relate to stress.
26. HSE Guidance Note HS(G)48 *Reducing Error and Influencing Behaviour* covers many of the causes of management stress.
27. There is a direct relationship between stress-related human failure and accidents at work.
28. Employers should pay attention to the various forms of human error and violations of workplace practices.
29. Strategies for managing stress must be considered at both organizational level and individual employee level.
30. The HSE Management Standards for Stress deal with a number of issues – the demands on people, the controls in place, the support available, relationships between managers and employees, the roles undertaken by people and the involvement of employees in organizational change.
31. There should be a formal system for communicating change within organizations.
32. Employers should create a healthy workplace with particular attention to environmental factors, cultural and social factors and lifestyle factors.
33. Health surveillance procedures should be directed at identifying stress and taking action to prevent or control stress.
34. Greater attention should be given to ergonomic factors, such as human characteristics, environmental factors, the man–machine interface and the total working system.
35. Employees benefit greatly from stress management programmes.
36. One of the main objectives of an occupational health scheme is the prevention or control of stress at work.
37. Employers need to recognize workplace stressors, measure and evaluate the levels of stress and install strategies for eliminating, controlling or reducing stress, together with systems for monitoring and reviewing progress in this area.

38. The Chartered Institute of Personnel and Development outline four main approaches to address stress at work, that is, policy, procedures and systems audit, a problem-centred approach, a well-being approach and an employee-centred approach.
39. Organizations should consider the introduction of corporate fitness programmes as a means of reducing stress.
40. The EU have laid down a number of principles for designing a stress prevention strategy incorporating promotion through improved design, participation of end users, better work organization, a holistic approach to the environment, an enabling organizational culture, attention to workers with special needs and economic feasibility.
41. Employers need to consider both their civil and criminal liabilities with respect to stress-induced injury.
42. The number of civil claims for stress-induced injury arising from work has increased dramatically since the landmark case of *Walker v Northumberland County Council* in 1994.
43. The following aspects are significant with respect to the civil liability of employers, namely the duty to take reasonable care, the employer's common law duty of care, the significance of negligence with respect to stress, and what is 'reasonably foreseeable injury'.
44. The Court of Appeal has produced guidelines with respect to employers' obligations for dealing with stress.
45. Recent civil claims are principally concerned with 'liability for psychiatric illness'.
46. Now that the HSE has taken stress at work as an important area for enforcement, it is conceivable that inspectors will be requiring evidence of measures taken by employers to prevent, control or reduce stress amongst employees. Failure to take appropriate measures, which would be based on the recent HSE guidance on stress, could result in the use of enforcement procedures, such as the service of an Improvement Notice under the Health and Safety at Work etc. Act 1974, to install these measures.
47. Risk assessment procedures under the Management of Health and Safety at Work Regulations must take into account the potential for stress amongst employees.
48. The preventive and protective measures arising from the risk assessment process in respect of stress must feature the organization's 'arrangements' for 'the effective planning, organization, control, monitoring and review' of these preventive and protective measures.
49. In the case of defined display screen equipment 'users', the potential for work-related stress must be considered in a workstation risk analysis.
50. The Disability Discrimination Act refers to disability arising from mental impairment, which could be the result of stress at work.
51. Employment tribunals are in a position to deal with actions arising from the Employment Rights Act 1996 in respect of bullying at work, constructive dismissal, discrimination, underpay and overwork, all of which can result in stress on employees.

Bibliography and further reading

Adams, J.D. (1980) *Understanding and Managing Stress. A Book of Readings.* University Associates: San Diego, CA.

Allport, G.W. (1961) *Pattern and Growth in Personality.* Holt, Rhinehart & Winston: New York.

Arbitration and Conciliation Advisory Service (1998) *ACAS Advisory Handbook Discipline at Work.* ACAS: London.

Arbitration and Conciliation Advisory Service (1998) *ACAS Code of Practice Disciplinary and Grievance Procedures in Employment.* ACAS: London.

Arbitration and Conciliation Advisory Service (2004) *Bullying and Harassment at Work. A Guide for Employers.* ACAS: London.

Bond, M. and Kilty, J. (1982) *Practical Methods of Dealing with Stress Human Research Project.* Department of Educational Studies, University of Surrey.

Cattell, R.B. (1946) *Personality.* McGraw-Hill: New York.

Chapman, A. (2002) *Hierarchy of Needs: 1990's Eight-Stage Model Based on Maslow.* www.businessballs.com.

Chartered Institute of Personnel and Development (2002) *A Corporate Strategy for Dealing with Stress at Work.* CIPD: London.

Cooper, C.L. (1998) *Theories of Organisational Stress.* Oxford University Press.

Cooper, C.L. and Marshall, J. (1978) Occupational sources of stress. A review of the literature relating to coronary heart disease and mental ill-health. *J Occup Psych* 49, 11–28.

Cooper, C.L. and Palmer, S. (2000) *Conquer your Stress.* CIPD: London.

Cox, T. (1993) *Stress Research and Stress Management. Putting Theory to Work CRR 61/1993.* HMSO: London.

Cox, T., Griffiths, A. and Barlow, C. (2000) *Organisational Interventions for Work Stress.* HSE Books: Sudbury.

Department of Employment (1976) Sex Discrimination Act 1976. HMSO: London.

Department of Employment (1996) Employment Protection Act 1996. HMSO: London.

Department of Employment (1996) Employment Rights Act 1996. HMSO: London.

Department of Employment (1998) Public Interest Disclosure Act 1998. HMSO: London.

Department of Trade and Industry (1998) Employment Equality Act 1998. HMSO: London.

Department of Trade and Industry (2003) Employment Equality (Religion or Belief) Regulations 2003. HMSO: London.

Department of Trade and Industry (2003) Employment Equality (Sexual Orientation) Regulations 2003. HMSO: London.

Earnshaw, J. and Cooper, C. (2001) *Stress and Employer Liability*. CIPD: London.

EC Health and Safety Directorate (1992) Recommendations on stress prevention strategies. Hygiea.

European Agency for Safety and Health at Work (2001) *Work-Related Stress: the European Picture*. EASHW: Brussels.

European Commission (1991) *Code of Practice on the Dignity of Men and Women at Work*. European Commission.

FIET (1992) *Recommendations on Limitations of Work-Related Stress and Pressure Affecting Salaried Employees*. FIET: London.

Fisher, J.M. (1999) *The Transition Curve: The Stages of Personal Transition*. Tenth International Personal Construct Congress, Berlin.

Green, M. (1982) *Practical Approaches to Meeting Business Needs for the Future*. AMED Henley Publications.

Health and Safety Commission (1975) *Health and Safety at Work etc. Act 1974*. HMSO: London.

Health and Safety Commission (1999) *Management of Health and Safety at Work Regulations 1999 and Approved Code of Practice*. HMSO: London.

Health and Safety Executive (1992) *Health and Safety (Display Screen Equipment) Regulations 1992 and Guidance*. HMSO: London.

Health and Safety Executive (1995) *Stress at Work: A Guide for Employers*. HSE Books: Sudbury.

Health and Safety Executive (1996) *Mental Health and Stress in the Workplace. A Guide for Employers*. HSE Books: Sudbury.

Health and Safety Executive (1997) *Violence at Work: A Guide for Employers INDG69L*. HSE Books: Sudbury.

Health and Safety Executive (1999) *Managing Stress at Work*. HSE Books: Sudbury.

Health and Safety Executive (1999) *Reducing Error and Influencing Behaviour HS(G)48*. HMSO: London.

Health and Safety Executive (2000) *Help on Work Related Stress: A Short Guide*. HSE Books: Sudbury.

Health and Safety Executive (2001) *An Assessment of Employee Assistance in Workplace Counselling Programmes in British Organisations*. HSE Research Report. HMSO: London.

Health and Safety Executive (2001) *Drug Misuse at Work. A Guide for Employers INDG91*. HSE Books: Sudbury.

Health and Safety Executive (2001) *Don't Mix It. A Guide for Employers on Alcohol at Work INDG240*. HSE Books: Sudbury.

Health and Safety Executive (2001) *Tackling Work-Related Stress. A Manager's Guide to Improving and Maintaining Employee Health and Well-being, HS(G)218*. HMSO: London.
Health and Safety Executive (2002) *Effective Teamwork Reducing the Psychosocial Risks: Case Studies in Practitioner Format CRR 393*. HMSO: London.
Health and Safety Executive (2002) *The Scale of Occupational Stress. A Further Analysis of the Impact of Demographic Factors and the Type of Job CRR 311*. HMSO: London.
Health and Safety Executive (2002) *The Scale of Occupational Stress. Bristol Stress and Health at Work Study CRR 265*. HMSO: London.
Health and Safety Executive (2002) *Violence at Work. New Findings from the 2000 British Crime Survey*. HSE Books: Sudbury.
Health and Safety Executive (2002) *Work Environment, Alcohol Consumption and Ill Health. The Whitehall Study CRR422*. HMSO: London.
Health and Safety Executive (2003) *Tackling Work-Related Stress. A Guide for Employers HSG218*. HSE Books: Sudbury.
Health and Safety Executive (2003) *Beacons of Excellence in Stress Prevention RR133*. HSE Books: Sudbury.
Health and Safety Executive (2003) *Best Practice in Rehabilitating Employees Following Absence Due to Work-related Stress RR138*. HSE Books: Sudbury.
Health and Safety Executive (2003) *Home Working Guidance for Employers and Employees on Health and Safety*. HSE Books: Sudbury.
Health and Safety Executive (2003) *Management Standards for Stress*. HSE Books: Sudbury.
Herzberg, F., Mansner, B. and Snyderman, B.B. (1959) *The Motivation to Work*. John Wiley: New York.
HMSO (1995) *Disability Discrimination Act 1995*. HMSO: London.
HMSO (1998) *Working Time Regulations 1998*. HMSO: London.
Holmes, D. and Rahe, J. (1967) The Social Readjustment Rating Scale. *J Psychosom Res* 11, 213–18.
Home Office (1976) *Race Relations Act 1976*. HMSO: London.
Home Office (2003) *Race Relations Act (Amendment) Regulations 2003*. HMSO: London.
International Labour Organisation (1974) *The Quality of Working Life*. ILO: Geneva.
Katz, D. and Brady, K. (1933) Racial stereotypes of one hundred college students. *J Abnorm Soc Psychol* 280–90.
Maslow, A.M. (1954) *Motivation and Personality*. Harper & Row: New York.
Mayo, E. (1952) *The Social Problems of an Industrial Civilisation*. Routledge & Kegan Paul.
Personnel Management (1988) Factsheet, 7 July.
Selye, H. (1976) *The Stress of Life*. McGraw-Hill: New York.
Smith, A.P., Brice, C., Collins, A., Matthews, V., and McNamara, R. (2000) *The Scale of Occupational Stress The Bristol Stress and Health at Work Study CRR 265*. HMSO: London.
Stranks, J. (1994) *Human Factors and Safety*. Pitman Publishing: London.
Stranks, J. (2003) *The Handbook of Health and Safety Practice*. Pearson Education: London.
Taylor, F.W. (1911) *Principles of Scientific Management*. Harper & Row: New York.
UMIST (1987) *Understanding Stress*. HMSO: London.

Index